Food Oils and
Their Uses

other AVI books

Food Oils

and Their Uses

THEODORE J. WEISS, Ph.D.

*Research Chemist, U.S. Department
of Agriculture, Agriculture Research
Service, Eastern Utilization Research
and Development Division, Dairy
Products Laboratory,
Washington, D.C.*

WESTPORT, CONNECTICUT

THE AVI PUBLISHING COMPANY, INC.

1970

Printed in the United States of America

BY MACK PRINTING COMPANY, EASTON, PENNSYLVANIA

Preface

A number of books have been written on the chemistry of fats and oils. They deal either with analytical methods or with chemical reactions and processing theory. There has been a tendency, however, for such books to treat formulation and utilization of finished products too lightly. They delve rather deeply into the scientific aspects of the field without considering the body of process methods known as "the art."

Much of the art appears in trade journal articles and patent literature. The scientific basis for information given in such sources is often mere conjecture or difficult-to-prove theory. This should not make the observations and techniques given less valid or of little importance.

It has been the author's experience that a great deal of interest exists in the food industry in obtaining technical information on fats and oils and products prepared from them. However, the student, food processor, plant operator, equipment manufacturer, etc., are often too busy with their own affairs to be able to wade through involved texts or to spend much time in contemplating technical literature.

The aim of this book is to provide abridged, technical information on fat and oil products and their uses. Material has been gathered from patents, trade journals, scientific journals, and personal experience. Practical aspects have been stressed without ignoring the theoretical. The book is directed at persons interested in the field whether or not they have been technically trained.

T. J. WEISS
New Orleans, La.

March 1970

Contents

CHAPTER 1

Chemical and Physical Properties of Fats and Oils

INTRODUCTION

Fats and oils make up one of three major classes of food materials, the others being carbohydrates and proteins. Fats and oils have been known since ancient times as they were easily isolated from their source. They found utility because of their unique physical properties. Fatty tissues from animal sources liberate free-floating fats on being boiled. Olives and sesame seeds yield oil on being pressed. Such fats and oils add flavor and lubricity to foods prepared with them.

The very commonness of fats and oils and the familiarity most people have with them has led to an ironic situation. The average user, whether a housewife, chef, baker, or food manufacturer, often has little real understanding of the character of fatty products. There are times when selection of the proper fat for a use situation can be very critical but is done incorrectly. Other times the selection need not be critical but a costly choice is made due to prejudice or to lack of adequate knowledge of the subject.

Much of the terminology used in fat and oil work developed over the years by processors who had only practical knowledge. Even the early chemists were hampered by incomplete understanding of their field of study. Therefore, a fat was defined as the oleaginous material which was solid at room temperature while the liquid form was called an oil.

The process of separating a fatty mass into a more liquid and a more solid fraction resulted in calling the former the oil and the latter the stearine portion. This led to the anomalous situation that oleo oil from beef fat was fairly solid at room temperature while the stearine separated from cottonseed oil could be liquid at a somewhat elevated room temperature.

COMPOSITION OF FATTY MATERIALS

Edible fats and oils are esters of the three carbon trihydric alcohol, glycerin, and various straight chained monocarboxylic acids known as fatty acids. The fatty acids of natural fats have 4 to 24 carbon atoms and, with minor exceptions, have an even number of such atoms. Figure 1 shows the structure of glycerin, fatty acids, and a fat derived from them.

The fatty acids may be saturated, monounsaturated, or polyunsaturated. Saturated acids have all of the hydrogen that the carbon chain will hold.

1

Table 1 lists them by chemical name, common name, and source. Mono-unsaturated acids have 2 hydrogen atoms missing, 1 from each of 2 adjoining carbon atoms, giving rise to 1 carbon–carbon double bond. By general agreement, polyunsaturated refers to a fatty acid with two or more double bonds.

Natural fatty acids usually exist in specific isomeric forms. Chemical isomers are compounds which occur in two or more forms although they

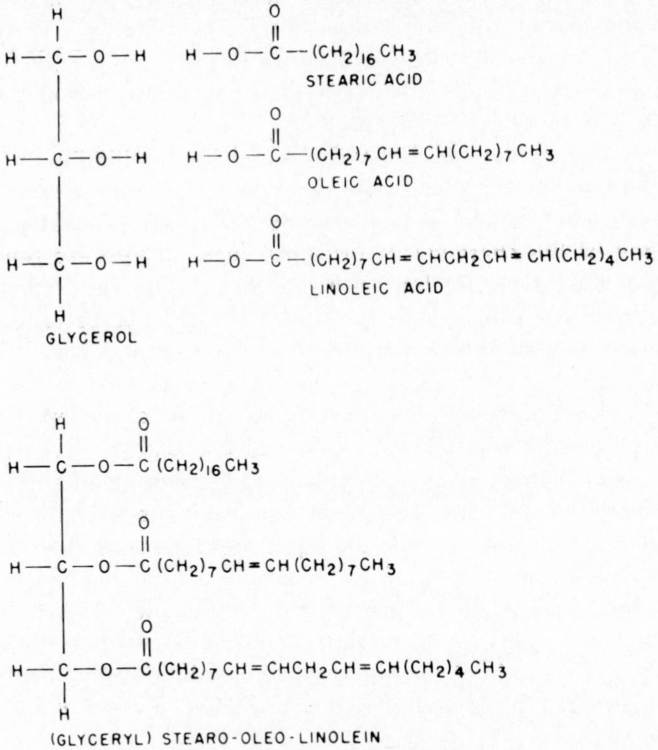

FIG. 1. STRUCTURAL FORMULA OF A TYPICAL TRIGLYCERIDE AND ITS COMPONENT PARTS

have the same number of carbon, hydrogen, and oxygen atoms. Unsaturated fatty acids can have positional isomers in relation to where the double bonds appear in the carbon chain. Natural fatty acids most frequently have the first double bond between the 9th and 10th carbons, the second between the 12th and 13th, and the third between the 15th and 16th carbons.

TABLE 1

SOME SATURATED FATTY ACIDS

Common Name	Chemical Name	No. C Atoms	Typical Source
Butyric	Butanoic	4	Butterfat
Caproic	Hexanoic	6	Butterfat
Caprylic	Octanoic	8	Coconut oil
Capric	Decanoic	10	Coconut oil
Lauric	Dodecanoic	12	Coconut oil
Myristic	Tetradecanoic	14	Coconut oil
Palmitic	Hexadecanoic	16	Most fats and oils
Stearic	Octadecanoic	18	Most fats and oils
Arachidic	Eicosanoic	20	Peanut oil
Behenic	Docosanoic	22	Peanut oil
Lignoceric	Tetracosanoic	24	Peanut oil

TABLE 2

SOME UNSATURATED FATTY ACIDS

Common Name	Chemical Name	No. C Atoms	No. Double Bonds	Typical Source
Caproleic	9-Decenoic	10	1	Butterfat
Lauroleic	9-Dodecenoic	12	1	Butterfat
Myristoleic	9-Tetradecenoic	14	1	Butterfat
Palmitoleic	9-Hexadecenoic	16	1	Animal fats
Oleic	9-Octadecenoic	18	1	Most fats and oils
Linoleic	9,12-Octadecadienoic	18	2	Most fats and oils
Linolenic	9,12,15-Octadecatrienoic	18	3	Soybean oil
Gadoleic	9-Eicosenoic	20	1	Fish oils
Erucic	13-Docosenoic	22	1	Rapeseed oil

Table 2 lists the various unsaturated fatty acids, their common names and sources. The most commonly occurring unsaturated fatty acids are oleic, linoleic, and linolenic.

Another type of isomer is the geometric isomer. Here the carbon chain is bent into a fixed position at each double bond. The carbon chain sections are either bent towards or away from each other. The former is called *cis* (meaning same side), the latter, *trans* (meaning across). Figure 2 illustrates these isomeric forms.

The natural acids are found most frequently in the *cis* form. *Trans* isomers are usually formed by chemical reactions, e.g., hydrogenation. *Cis* acids have a considerably lower melting point than *trans* acids of the same chain length. Thus, oleic acid (*cis* octadecenoic) melts at 16°C and is therefore liquid at room temperature. Elaidic acid (*trans* octadecenoic) has a melting point of 44°C and is solid at room temperature. Infrared spectroscopy is used for analysis of *trans* isomers in fats.

The length of the carbon chain also affects melting point. Melting point increases as the chain becomes longer. It is interesting to note that stearic acid, the saturated 18 carbon acid, melts at 69°C. Oleic acid has a melting point close to that of caprylic acid. The former has 18 carbons and one *cis* double bond. The latter is saturated with 8 carbons. Elaidic acid has 18 carbons and one *trans* double bond. It has a melting point close to that of lauric acid with a 12 carbon chain.

The fatty acids are found in natural fats as complex mixtures. There are three positions available on the glycerin molecule for the esterification of the fatty acids. The physical characteristics of the various triglyceride fats depend on the type, quantity and distribution of the acids on the glycerin molecule. A simple illustration of this was demonstrated by Nor-

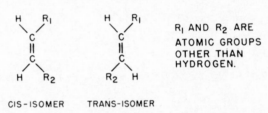

CIS-ISOMER   TRANS-ISOMER

Fig. 2. Structural Model of cis-trans Isomers

ris and Mattil (1946). They randomized mixtures of triolein and tripalmitin by interesterification and analyzed the glyceride composition of the final products.

Figure 3 shows the theoretical composition of one random mixture. It should be noted that a distinction has been made between the 1- and 2-positions on the glycerin molecule. The characteristics of a triglyceride depend on which position each fatty acid occurs. This is most important in the sharpness of melting of cocoa butter and in the crystal structure of lard. It is of lesser importance with other types of fats and oils.

The melting point of fatty mixtures changes on randomization. A mixture of 30% tripalmitin and 70% triolein had a melting point of 136°F before interesterification and 119°F after the reaction (Norris and Mattil 1946).

There are also a number of minor components incorporated in natural fats. Simplest of these are mono- and diglycerides. Glycerin is esterified with only 1 or 2 fatty acids, leaving free hydrophilic hydroxyl groups. This makes the monoglyceride an important emulsifier.

Phospholipids are also important as emulsifiers. They are compounds containing fatty acids and phosphoric acid among other chemical entities.

Lecithin is a glycerin ester of two fatty acids and phosphoric acid combined with choline, a nitrogenous compound. Cephalin is similar with ethanolamine, a different nitrogen-containing compound, replacing choline. Inositol phosphatides are compounds of inositol, phosphoric acid, fatty acids, ethanolamine, tartaric acid, and sugars.

Soybean oil phospholipids consist of 29% lecithin, 31% cephalin, and 40% inositol phosphatides. They are commonly called "lecithin." Com-

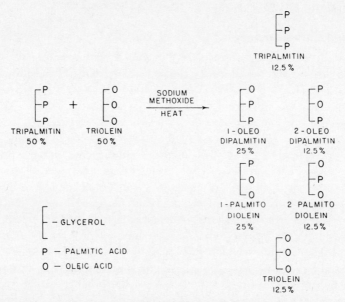

FIG. 3. RANDOMIZED MIXTURE OF TRIPALMITIN AND TRIOLEIN

mercial lecithin is obtained primarily from soybean oil. Some are also available from corn oil and egg yolk.

Sterols are complex polycyclic alcohols. They are found as a class of compounds in all oils. Cholesterol is found only in animal fats. Plant sterols are similar in composition but sufficiently different to suppress absorption of dietary cholesterol by the body. Sterols may have some effect as emulsifiers. Joffe (1942) has postulated that cholesterol in egg yolk aids in the formation of the mayonnaise emulsion although lecithin and lecithoproteins are considered to be the major emulsifiers in yolk.

Tocopherols are natural antioxidants found in vegetable oils. They are also known as vitamin E. They are cyclic compounds with a hydroxyl group and heterocyclic oxygen. Tocopherols are easily oxidized to quinones. They, therefore, have natural antioxidant activity.

EVALUATION OF PROPERTIES OF FATS

A large number of test methods have been applied to the evaluation of fats and oils (Mehlenbacher 1960; American Oil Chemists' Society 1967). The methods are mostly empirical in nature. In effect, a sample of oil is treated in a certain reproducible manner and the results are given in numbers which have meaning only in terms defined by the test. A few methods can be chemically more exact in their evaluation. The analytical chemist is trying to expand the list of precise methods. However, the nature of fats and oils makes the task difficult.

Gas-liquid Chromatography

Gas-liquid chromatography (GLC) is one of the more recent developments which is used to determine the exact fatty acid composition of fats and oils and related products (AOCS Method Cc 1-62). In this method, the fatty acids are converted from glyceride to methyl esters. They are then injected into the heated chamber of the GLC apparatus. The volatilized esters are picked up in a stream of inert gas such as helium. They are passed through a 2.5- to 6-ft length of tubing containing a polyester or other adsorbent material. The various component methyl esters are separated into distinct zones of pure ester. A detector notes the presence of each ester in the stream of inert gas as it appears and signals a recorder. The recorder indicates the signal as a peak on a travelling chart. The area under the peak is related to the amount of ester. The position of the peak determines which fatty acid is present. This must first be determined by passing pure known esters through the system and measuring the distance of the peak detected from the point at which the ester was injected into the system. A complete analysis can be made with a fraction of a gram of triglyceride. Many refinements of the basic apparatus and techniques are currently available.

Iodine Value

Iodine value, also referred to as iodine number, is defined as the number of centigrams iodine absorbed by one gram of fat or percent iodine absorbed (AOCS Method Cd 1-25). The test is empirical in that the theoretical amount of iodine is never absorbed. The various iodine reagents contain chlorine or bromine as accelerators, but even with these, the iodine absorption merely approaches completeness. The test must be run very precisely and timed carefully to be reproducible.

Iodine value actually corresponds to the average number of double bonds in the fat measured. It does not give the distribution of double bonds among the fatty acids present. Therefore, it does not tell the analyst what fatty acids are present in the fat.

TABLE 3

IODINE VALUES OF VARIOUS UNSATURATED FATTY ACIDS AND
THEIR TRIGLYCERIDES

Fatty Acid	No. C Atoms	No. Double Bonds	Iodine Value	
			Fatty Acid	Triglyceride
Decenoic	10	1	149.1	138.77
Lauroleic	12	1	128.0	120.32
Myristoleic	14	1	112.1	106.20
Palmitoleic	16	1	99.78	95.04
Oleic	18	1	89.87	86.01
Linoleic	18	2	181.04	173.21
Linolenic	18	3	273.52	261.61
Eicosenoic	20	1	81.75	78.54
Erucic	22	1	74.98	72.27

Source: Bailey (1950).

It is becoming common practice among fat and oil research workers to obtain a fatty acid composition by GLC and to calculate the iodine value from this composition. Table 3 gives iodine values of the common unsaturated fatty acids as such and as glyceride esters.

The iodine value method is still used by plant control laboratories. GLC apparatus is too elaborate and expensive for general use. Iodine value is a guidance tool for purchase of raw materials and for control of hydrogenation.

Purchase specifications for shortening products often include iodine value in order to define the product. This is satisfactory as long as the same raw materials are used from batch to batch and the shortening comes from the same refinery. With these factors being constant, iodine value will correlate with hardness and oxidative stability of the shortening.

Hydrogenated cottonseed oil has a lower iodine value than hydrogenated soybean oil for an equivalent hardness. Temperature, pressure, and catalyst activity in hydrogenation affect the relationship between iodine value and stability, crystal structure, and hardness of a finished shortening. Therefore, iodine value can not be used by itself to describe a shortening product.

Most processors have developed their hydrogenation procedures over the years. Since no two processors have the same practices or equipment, it would be most unlikely for them to have a similar iodine value for equivalent competitive products. Thus, iodine value becomes useful primarily for the processor and should be used with caution by the purchaser of shortenings.

Saponification Value

Saponification value has been completely superseded by GLC analysis. It is defined as the weight in milligrams of potassium hydroxide needed to completely saponify 1 gm of fat (AOCS Method Cd 3-25). Saponification value is related inversely to the average molecular weight of the fat. As with iodine value, it does not give the exact fatty acid composition.

Melting Point

Melting point is a rather broad, general term when applied to fats and oils. It is not even a physically correct term when compared to the evaluation of pure compounds. Pure compounds usually melt quite sharply. Most fats are mixtures of triglyceride esters of varying degrees of unsaturation. The fat is considered to be completely melted when, in fact, the harder components are merely dissolved in the softer components. "Solution point" would be a more accurate term. While this may sound like quibbling, the intention is to give a true picture of what happens when a fat is heated. It will make discussions of the various functions involving "melting" more understandable.

FAC Melting Point.—There are actually several melting points involved in fat and oil work. The FAC (Fat Analysis Committee of the American Oil Chemists' Society) melting point is also called the "closed capillary" melting point (AOCS Method Cc 1-25). It is the same procedure as is used in organic chemistry. The sample of fat is placed in a glass capillary tube. The end is sealed by fusing the glass. The tube of fat is then put in a refrigerator at 40°–50°F for several hours to set up completely. The tube is then heated slowly in a water bath. The melting point is the temperature at which the fat becomes completely clear.

The chilling of the fat sample for several hours prior to the determination of this or any other melting point is necessitated by an important phenomenon in triglyceride chemistry. Oils are viscous compounds, especially at low temperatures. The molecules are quite sluggish and are slow to arrange themselves in the proper alignment for crystallization. A fat which has been solidified for too short a time will not have formed a sufficiently stable crystal to give a reproducible melting point.

Solidification Point.—A corollary to melting point is found in the determination of solidification of a fat. A lower temperature is required to cause solidification than to bring about melting. Orientation of the sluggish fat molecules takes longer than their disorientation. Slow orientation leads to supercooling in which the fat remains in liquid form well below its melting point. Lard is an outstanding example of a fat which can remain supercooled for a long time, sometimes for hours.

On the other hand, a fat which has been just barely melted will solidify quite rapidly on being cooled only a few degrees below its melting point. The molecules are still oriented in the proper alignment for crystallization. It takes several hours at temperatures a few degrees above melting or at a considerably higher temperature for a short time in order to disrupt the precrystalline molecular aggregates.

Wiley Melting Point.—A second and more widely used melting point was developed by Wiley (AOCS Method Cc 2-38). In this method, a round disc of fat is cast in a mold and chilled for 2 hr at refrigerator temperature. The sample is then suspended in the center of an alcohol-water bath and heated slowly until it changes shape. The difficulty in this method is the decision on what the proper shape is at the end point and how to judge it. Some analysts look for a sphere (which is official), others for a football shape. In one case, the end point is judged as being the temperature at which the inside of the fat disc is observed to turn over. This makes it difficult to obtain good agreement between analysts on the Wiley melting point of any given sample.

Since the fat is not completely melted at the Wiley melting point, it is obvious that the FAC melting point is higher. The difference between the two melting points is the smallest with sharply melting fats such as confectioners' hard butters and the largest with wide plastic range shortenings.

Softening Point.—The softening point (AOCS Method Cc 3-25) is also known as the open capillary melting point. It is not used frequently. In this method, the fat sample is placed in a capillary tube similar to that used for the FAC melting point. Here, however, the tube is open at both ends. The fat is chilled and then placed in a water bath as with the closed capillary melting point method. The end point is taken as the temperature at which the fat rises a definite amount in the tube. This temperature is below the Wiley melting point to about the same extent that the Wiley is below the FAC melting point.

Dilatometry

There are other methods which depend on the melting or solidification of solid triglyceride components in the liquid glyceride portions of the entire fat mass. The older methods gave partial information and may still find special usage in specific areas of fat processing. They have been supplanted by dilatometric methods which depend on the change in volume of a solid fat upon melting.

Dilatometry was thoroughly explored by Hofgaard, Singleton, and Bailey (Bailey 1950). Their method was a long and arduous procedure which is rarely used today and then only as a precise research tool. It should be discussed, however, in order to better understand and appreciate rapid empirical methods that are so important in modern fat technology.

It should first be understood that most seemingly solid fats actually consist of mixtures of solid fat components forming a crystal matrix. This holds the liquid oil portion in much the same manner that a sponge holds water. If the fat is chilled to a sufficiently low temperature, such as —30°C, it will contain 100% solid fats. Above the FAC melting point, the same material will be a completely liquid oil with no solid fats present.

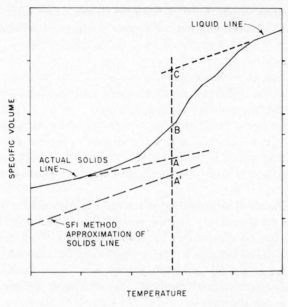

From Bailey (1950)

FIG. 4. DILATOMETRIC CURVE OF TYPICAL FAT

Relation between specific volume and temperature is shown with solid and liquid lines extrapolated for calculation of percent solid fats. SFI method solids line is included for comparison of methods.

Pure chemical compounds melt sharply, going from a complete solid to a complete liquid within a fraction of a degree rise in temperature. Fats may be relatively pure triglycerides as a group of compounds. However, the individual triglycerides are highly variable in physical characteristics depending on their particular fatty acid makeup. In one sense, a single triglyceride molecular type can be considered as an impurity for another single but different triglyceride type. In another sense, one triglyceride can and does act as a solvent for another triglyceride.

When a normal fat is heated from a completely solid state, it does not melt as a single pure compound but as the mixture that it really is. The entire mixture melts by degrees as its temperature is increased, the most unsaturated fats melting first, the most saturated melting last. Solid fat components expand with temperature increase at a different rate than do liquid fat components, but the greatest expansion takes place when any particular solid fat portion becomes liquid.

Figure 4 shows a hypothetical dilatometric curve. The plot of specific volumes versus temperature is used to calculate the percent solid fats at

TABLE 4

RELATIONSHIP BETWEEN SOLID FAT INDEX AND TRUE PERCENT
SOLID FATS IN SHORTENING

Temperature, °C	% Solids	SFI
50	0.0	0.0
45	3.3	2.9
40	6.8	5.7
35	10.9	9.4
30	15.4	12.9
25	17.2	14.0
20	20.6	16.7
15	29.6	21.7
10	39.8	27.8

Source: Bailey (1950).

any given temperature. This is done by extrapolating both the solid and the liquid curves and determining the relative position of the actual specific volume curve at the temperature concerned. The percent solids is equal to the unmelted fractional portion of the whole mass. Thus in Fig. 4:

$$\text{Percent solid fats} = BC/AC$$

In the original method, the solids were determined for various fats at intervals of $1°–2°C$ from the completely solid to the completely liquified sample. This was modified into a relatively rapid control method as well as empirical research tool. Obviously, certain assumptions and approximations would have to be made, some of which were admittedly but justifiably inaccurate. This led to the term Solid Fat Index (SFI) which was related to but more or less different from true percent solid fats (AOCS Method Cd 10-57) (Fulton et al. 1954).

Figure 4 includes a hypothetical SFI plot which can be compared with the equivalent true dilatometric curve. In this calculation the difference

$$\text{SFI} = BC/A'C$$

is simply that the SFI method assumes rather than measures the solid fat expansion line. The temperatures required for measurement of the true solids line are very low and difficult to control. The SFI solids line is given the same slope as the liquid line which is easy to measure. It is placed 0.100 specific volume units below the liquid line. This is a convenient approximation. True values vary from it by different amounts depending on the actual composition of the fat being measured. Table 4 shows a comparison between percent solid fats and SFI values for a typical shortening.

In addition to the aforementioned approximations, the SFI determination has been further simplified by selection of specific temperatures at which readings are to be made. These are usually 50°, 70°, 80°, 92°, 100°, and 104°F. Individual processors use either 100° or 104°F, but rarely both. Margarine manufacturers have generally agreed to use SFI

TABLE 5

SFI VALUES OF NATURAL FATS

Fat	Melting Point, °F	SFI Value				
		50°F	70°F	80°F	92°F	100°F
Butter	97	32	12	9	3	0
Cocoa butter	85	62	48	8	0	0
Coconut oil	79	55	27	0	0	0
Lard	110	25	20	12	4	2
Palm oil	103	34	12	9	6	4
Palm kernel oil	84	49	33	13	0	0
Tallow	118	39	30	28	23	18

Source: Weiss (1963).

values at 50°, 70°, and 92°F. Some processors define a product using only 1 or 2 points on the SFI curve. This can be misleading since it takes values at a minimum of three temperatures to adequately describe the physical characteristics of a fat. This is especially true in control of hydrogenation of oils. If some factor, such as residual soap in refined oil, reduces selectivity of hydrogenation, it will be readily detected by a three-point SFI curve.

The significance of SFI in various aspects of fat and oil production will be discussed in later chapters. Table 5 shows the SFI values for various natural fats.

Analysis of Solidification Phenomena

Congeal Point.—Some special methods are based on solidification of fats. Congeal point, also known as setting point, is one such method (AOCS Method Cc 14-59). Here a melted sample of fat is supercooled under closely defined conditions until it begins to harden. The heat released by crystallization warms the sample to a specific temperature depending on the type and amount of solid fats formed. The maximum temperature reached is the congeal point. The technique is accurate for determining the melting point of pure compounds. However, it is subject to considerable error when used with normal fats.

Titer.—The method used for determining congeal point of fats is also used for measuring the melting point of fatty acids. In this case it is known as titer determination (AOCS Method Cc 12-59). The method has been used primarily for evaluation of fatty acids for the manufacture of soap. Titer has little significance in edible oil processing. Its major function is in describing the hardness of fully hydrogenated fats and oils.

The melting point of a triglyceride and its titer are considerably different with liquid oils. Both values increase on hydrogenation of the oil, becoming almost identical when complete saturation is reached. The FAC melting point of a hardfat can be used in place of titer to define its hardness.

Cold Test.—Cold test is used to determine the ability of a salad oil to withstand refrigerator storage. An oil is considered to pass the minimum test if it remains perfectly clear after standing 5.5 hr in an ice bath (AOCS Method Cc 11-53). Even a trace of fat crystal in the oil is judged as indicating the end point. In actual practice the length of time it takes for a slight cloud to first appear in the oil is reported as the cold test. A 20-hr oil is considered to be quite good.

The cold test method was developed to evaluate cottonseed oil for the production of mayonnaise. If mayonnaise were to be prepared from an oil which solidified in the refrigerator, the emulsion would break and the mayonnaise would be spoiled.

Many times an oil will develop a slight cloud in relatively few hours but never progress beyond that stage. This is especially true of unhydrogenated soybean oil which may have become contaminated with hydrogenated fats. Mayonnaise prepared from such oil will be quite stable under refrigeration. On the other hand, a poorly winterized cottonseed oil with crystal inhibitors added may pass a minimum cold test but develop a voluminous mass of crystal fats some time after the crystals begin to precipitate. Mayonnaise made with an oil of this type could break down under refrigeration.

The cold test is a slow procedure. Salad oil is often winterized and pumped into storage tanks before the test is completed. A more rapid and meaningful test has been developed for the evaluation of salad oil. The method calls for chilling the sample in a bath at $-60°C$ for about 15 min and placing it in another bath at refrigerator temperature, e.g.,

10°C. If the sample contains solid fats after 30 min at this temperature, it is considered as having failed the test (Pohle 1969).

Crystal Structure

The crystal structure of a solid fat is critical in some areas of fat utilization (Hoerr and Ziemba 1965). Crystal structure may be observed by X-ray diffraction of the fat. In this method, the sample to be observed is allowed to solidify in a hole drilled through a metal disc of special thickness. The disc is placed in an X-ray beam. The X-rays are deflected or diffracted by the fat into a series of concentric cones which impinge on a photographic plate placed a known distance from the sample. The developed plate consists of a series of rings, the diameters of which are translated by the analyst into molecular spacings. These are the distances between the triglyceride molecules in the fatty crystal under examination.

A number of crystal forms have been identified and designated by the letters of the Greek alphabet. Each form has a specific molecular dimen-

TABLE 6

CRYSTAL STRUCTURE OF VARIOUS FULLY HYDROGENATED FATS IN
THEIR MOST STABLE STATE

Alpha	Beta-prime	Beta
Acetoglycerides	Cottonseed	Soybean
	Palm	Safflower
	Rapeseed	Sunflower
	Herring	Sesame
	Menhaden	Peanut
	Whale	Corn
	Sardine	Olive
	Tallow	Coconut
	Butter	Palm kernel
		Lard
		Cocoa butter

Source: Wiedermann (1968).

sion. While various subtypes have also been recognized, the major types are the only ones of real concern in shortening performance evaluation. Fats and the crystal form they assume when in their most stable state are given in Table 6. Physically, fats in the alpha form are waxy, in the beta-prime, fine grained, and in the beta, coarsely crystalline.

Fats exhibit polymorphism, meaning that they can exist in different crystal forms in varying degrees of stability. Cocoa butter offers an extreme example of polymorphism in fats (DuRoss and Knightly 1965). Rapid chilling of cocoa butter on dry ice gives the gamma form which is very unstable and converts in seconds to the alpha crystal. As the cocoa butter warms it undergoes transition to the beta-prime form in about 1 hr. It can remain in this form for 30 days before transforming to the final and most stable beta crystal. In the processing of chocolate the molten mass is chilled and tempered with great care in order to produce finished chocolate in its most stable crystalline structure.

In some cases, a fatty product can be formulated and produced in an unstable form which it will retain for weeks as long as the storage temperature is relatively low. This has happened with margarine which was

FIG. 5. PHOTOMICROGRAPHS OF TYPICAL ALPHA, BETA, AND BETA-PRIME FAT CRYSTALS (400×)

Alpha crystals (left): Acetylated monoglycerides of fully hydro-genated lard. Beta crystals (center): Fully hydrogenated peanut oil in liquid oil. Beta-prime crystals (right): Fully hydrogenated cottonseed oil in liquid oil.

found to become grainy on long storage. The formulation was one which was later found to have beta crystals in its most stable state (Merker et al. 1958).

Another method for observing crystal structure is by direct examination under the microscope. A drop of melted fat is placed on a glass slide under a cover slip and allowed to cool. It is then held for a few hours at room temperature (or close to its melting point for fully hydrogenated fats). The crystals are then observed under polarized light at about 400× magnification. Figure 5 shows photomicrographs of typical crystal forms.

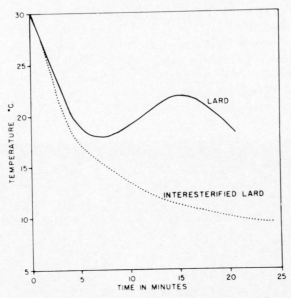

FIG. 6.   MANUALLY OBTAINED COOLING CURVES FOR LARD
AND INTERESTERIFIED LARD

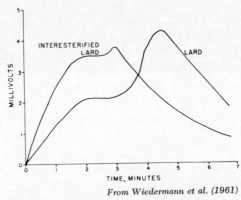

From Wiedermann et al. (1961)

FIG. 7.   MECHANICAL DIFFERENTIAL COOLING
CURVES FOR LARD AND INTERESTERIFIED LARD

As a general rule, fats which are made up of fairly uniform triglyceride
molecules have beta crystals in their most stable state. This seems so
because such triglyceride molecules can deposit easily without interrup-
tion and form large crystals. Fats which are stable in the beta-prime
form contain a mixture of types of triglyceride molecules which seems to
prevent them from growing too large. The mixed molecules do not

deposit on the crystal surface in as orderly a manner as do unmixed molecules. Soybean, peanut, and corn oils, for example, have low levels of palmitic acid (less than 12%). When fully hydrogenated, these fats are essentially all tristearin. They form stable beta crystals. Cottonseed oil with over 20% and palm oil with about 50% palmitic acid have mixtures of stearic and palmitic acids in their fully hydrogenated molecules. They form small beta-prime crystals. Rapeseed and marine oils have high levels of 20 and 22 carbon fatty acids which act in the same manner as palmitic acid in affecting crystal structure.

Lard and cocoa butter each contain about 25% palmitic acid. Still they crystallize in beta form. This is due to the fact that the triglycerides of these fats are fairly uniform in composition. Palmitic acid in lard is always found on the second carbon of the glycerin moiety. Cocoa butter has its oleic acid in the same position (Wiedermann *et al.* 1961).

Thermal Analysis

Cooling Curve.—Thermal analysis is somewhat related to crystal structure. The basis of the cooling curve is that crystals of fat give off heat on solidifying from liquid oils and absorb heat on melting (Jacobson *et al.* 1961). Large beta crystals give up heat so rapidly during their formation, that the temperature of the fat may rise rapidly during the chilling operation. Figure 6 shows typical cooling curves obtained manually by allowing a sample of fat to cool in a titer tube inserted in an ice bath and recording the temperatures obtained at regular intervals (Quimby *et al.* 1953).

Figure 7 shows mechanically drawn differential cooling curves (Jacobson *et al.* 1961). These are obtained by recording the difference between the electric current derived from a pair of matched thermocouples which have been inserted in a pair of matched test tubes. One tube contains the sample to be measured. The other tube is filled with an oil which will not solidify in an ice bath. The two tubes are first heated in boiling water, then cooled in an ice bath. The difference in current correlates with the difference in temperature between the two tubes. The curves in Fig. 7 demonstrate that the rate of heat evolution from the formation of beta-prime crystals (interesterified lard) is relatively low. Beta crystal formation (lard) results in a sudden burst of heat and subsequent rise in temperature (Wiedermann *et al.* 1961).

Heating Curves.—Heating curves show the absorption of heat by a fat as it is warmed and melted. They are not, as it might seem, the opposite of cooling curves. A cooling curve is run rapidly with the sample crystallizing in a characteristic metastable state. Heating curves can be determined for fats pretempered into specific crystal forms. This is illustrated by curves obtained for cocoa butter by Johnston and Price (see Chap. 12).

Consistency

Finished shortening and margarine products which have been chilled on a Votator or other plasticizing machine are evaluated for firmness by various penetration methods. One of these uses the ASTM grease pene-

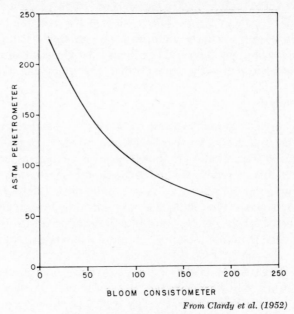

BLOOM CONSISTOMETER

From Clardy et al. (1952)

Fig. 8.   Relationship Between ASTM Penetrometer
and Bloom Consistometer Values

trometer (AOCS Method Cc 16-60). It measures the depth to which a standard cone penetrates into the surface of the shortening on being dropped onto it.

A second instrument is the Bloom consistometer (Bloom 1925). It measures the resistance of the shortening to being penetrated by a standard ring being pushed into it. The scale units are arbitrary but could easily be registered as pounds of weight required to force the ring into the sample. The Bloom method requires a practiced arm movement to achieve a uniform rate of penetration and thereby reproducible values.

The main advantage of the Bloom method is that the instrument can be carried to the sample in the warehouse with the determination usually being made directly in the package. The ASTM method requires removal of a sample from the package and usually also from the warehouse to a constant temperature room. Figure 8 shows the correlation between ASTM penetration and Bloom consistency for a typical shortening.

SFI also correlates with consistency since the solid fat content of the shortening is the main contributing factor to its solidity. This correlation is given in Fig. 9. It is not perfect, however, as crystal structure enters

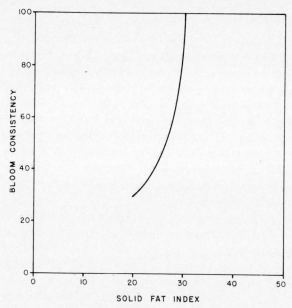

Fig. 9. Relationship Between Bloom Consistometer Values and SFI of a Shortening

into the overall relationship between solid fats and texture. SFI is an empirical method and deviates from true percent solid fat content depending partly on the crystal structure of the fat concerned. More to the point, a fine textured beta-prime crystal is more finely dispersed than a coarse beta crystal. A beta-prime fat creates a stiffer texture for a given weight of fat solids than does a beta type fat. Lard chilled into a temporary beta-prime phase softens when it is transformed on standing into its permanent grainy structure. Vegetable shortening has been known to undergo such transition when the base oil was improperly hydrogenated or when it was formulated with a beta crystalline hardfat.

Consistency of a finished shortening is also affected by chilling and tempering conditions. This is related to the size and distribution of fat crystals in the shortening.

DETERMINATION OF QUALITY

The quality of a shortening or other fatty product is independent of its physical characteristics. Quality is measured in a number of ways but invariably reflects what might be considered as the purity of a product. Obviously, not all products can be judged by the same standards.

Sensory Evaluation

The simplest quality test would seem to be organoleptic evaluation, that is, the determination of flavor and odor. It is really not simple. The detection of flavor and odor of oil products requires experience. Inexperienced people are often unable to detect odors and flavors which the trained and experienced taster would judge to be quite strong. Even with experience, the sensitivity of the taster will vary from day to day or with time of day due to a number of factors. The solution to this problem is to rely on a panel of trained tasters, preferably at least six in number. The proper operation of sensory evaluation panels and factors influencing them has been the subject of a number of books and articles. These have been reviewed by Ellis (1967).

Most oils are deodorized to complete blandness before being used in the production of shortenings or margarine. On storage, these oils begin to develop characteristic flavors and odors which have been termed reverted. By definition, reversion is the return of the flavor that the particular oil had before it was deodorized. Ordinary chemical methods have not been able to detect reverted flavors in oils although some workers, primarily Smouse and Chang (1967), have been able to isolate and identify 71 compounds which have been formed during reversion of soybean oil.

Chemical Tests

Peroxide Value.—When an oil becomes oxidized, it first develops hydroperoxides. The extent of peroxidation is measured by the amount of free iodine which the oxidized fat can liberate from potassium iodide. The results are expressed as peroxide value, the milliequivalents of iodine formed per kilogram of fat (AOCS Method Cd 8-53).

Hydroperoxides have no flavor or odor. However, they do break down rapidly to form aldehydes which have a strong, disagreeable flavor and odor. Peroxide value correlates to some extent with the off-flavor caused by aldehydes and other oxidation products. The overall flavor defect is called oxidative rancidity. A peroxide value of 1 me/kg is borderline in terms of rancidity. Some fat samples have been judged to be rancid with a peroxide value of 0.5. Other samples have no apparent rancidity with a peroxide value of two. A freshly deodorized fat will have a zero peroxide value. This will rise, however, when the fat is shipped. This is especially so with bulk shipments in tank wagons or tank cars as the product is loaded and unloaded in the molten state.

Active Oxygen Method.—Peroxide value is used to determine the end point in the Swift Active Oxygen Method (AOM), a technique for measuring fat stability (AOCS Method Cd 12-57). To determine AOM, the sample to be analyzed is heated in a test tube in a water bath at 97.8°C, while air is blown through the sample. The fat oxidizes, slowly at first, then rapidly. Extent of oxidation is measured by the peroxide value. The AOM value is expressed as the number of hours required for the sample to reach a peroxide value of 100 me/kg.

The AOM method was designed to evaluate unstabilized lard, the shelf-life of which does correlate well with AOM. The use of stabilizers (antioxidants) has lessened the accuracy of predicting shelf-life by AOM. The development of vegetable oils and hydrogenated shortenings has made AOM obsolescent. Other methods have been developed to improve our ability to predict actual shelf-life of fats and oils.

Oxygen Bomb.—The oxygen bomb method was first designed to speed up evaluation of the potential stability of a shortening. It was judged to be 1.4 times faster than AOM and 40–50 times faster than the Schaal Oven Test (Gearhart *et al.* 1957).

In the bomb test, the sample to be analyzed is sealed in a heavy-walled metal container or bomb attached to a pressure recorder. Oxygen is admitted to the bomb at 100 psi. The bomb is then placed in a boiling water bath. The sample absorbs oxygen slowly. The end point is reached when the sample suddenly begins to absorb oxygen rapidly. This is noted by the pressure recorder as a sharp drop in internal pressure of the bomb.

It has been found that the bomb method results correlate more closely with the Schaal Oven Test and with shelf-life than does AOM (Pohle *et al.* 1964). In addition, the oxygen bomb can be used for determining stability of foods which contain fats, e.g., potato chips and crackers. It is not necessary to extract the fat from the food material for the determination.

Schaal Oven Test.—The Schaal Oven Test has been found to correlate well with shelf-life (Joyner and McIntyre 1938). It is an old method which has many modifications that have not been clearly defined. Basically, it is an accelerated storage test in which the sample to be evaluated, either fat or fatty food, is held at a relatively high temperature for a length of time and then examined for signs of rancidity. The storage temperature is usually 140°F but can be higher or lower. The test can be run for a specific length of time with the product being considered as satisfactory if it is judged to be free of rancidity. The product can be held until it is rancid and its shelf-life be predicted on the basis of the length of time required to develop rancidity. Judgment of rancidity is the major difficulty. A sensory evaluation panel is probably the best tool for such judgment.

Free Fatty Acid.—Free fatty acid (FFA) content of a fat is a good indicator of overall quality. It is determined by titration of the sample with a standard solution of sodium hydroxide (AOCS Method Ca 5a-40). The results are calculated as free oleic acid. Since fats are esters of glycerin and fatty acids, FFA is the result of hydrolysis of the fats. Moisture must be present for hydrolysis to take place.

TABLE 7

EFFECT OF FREE FATTY ACID ON SMOKE POINT OF SHORTENINGS

Free Fatty Acid, %	Smoke Point, °F	Flash Point, °F	Fire Point, °F
0.04	425	620	690
0.06	410		
0.08	400		
0.10	390	595	685
0.20	375		
0.40	350		
0.60	340		
0.80	330		
1.00	320	585	680

Source: Swern (1964).

Hydrolysis is greatly accelerated by the enzyme, lipase. Crude vegetable oils may have excessive FFA if they have been extracted from improperly stored seed. The seed lipases are usually practically inactive unless the seed becomes wet and begins to germinate. Lard and tallow are rendered from fatty animal tissues. If the tissues are allowed to stand for too long a time before being rendered, the lipases will cause development of high FFA. Undeodorized meat fats should have a maximum of 1% FFA.

Vegetable oils are alkali refined to remove FFA. After refining, oil is often held in storage tanks. Water vapor can get into these tanks through breather pipes and condense on cooling. FFA can increase again because of the presence of such water. Deodorization, the final process before the shortening is packaged, lowers FFA to 0.05% or less. However, moisture in pipe lines, feed tanks, and tank wagons and cars can cause subsequent increase in FFA.

Smoke Point.—Smoke point is the temperature at which a fat will begin to emanate continuous wisps of smoke (AOCS Method Cc 9a-48). FFA content of the fat correlates with smoke point. This is given in Table 7. Flash and fire points are also given. It should be noted that extremely high temperatures are required for a fat to actually ignite.

Other materials in a fat can also lower the smoke point. Monoglycerides are present in all fats. A good frying shortening will have less than 0.4% monoglycerides and less than 0.05% FFA. Its smoke point will be 425°F or higher. This is 35°F above the maximum frying temperatures normally used. Addition of monoglycerides, as in the preparation of all-purpose household shortenings, lowers the smoke point to about 375°F. Proteinaceous residues in undeodorized lard and tallow can lower the smoke point to about 325°F. Residues from foods being fried in fats also lower the smoke point so that a high smoke point frying fat does not remain so for very long once frying in it has begun.

Color

Oil color is another indicator of quality. Exact color is specific for each type of oil. Some oils are naturally darker than others. Certain pigments are found in particular oils which behave differently from others. Color is removed by one of a number of processes called bleaching. The various pigments found in different oils react differently to any given process.

Light colored oils and white shortenings are preferred by the consumer as a general rule. However, olive oil is expected to be green. Oil used for pourable salad dressings is preferably dark, even to the extent of being artificially colored with carotenoid pigments.

If an oil is dark when it is supposed to be light, it may be due to one of several reasons. The oil may have been processed from a poor grade of crude oil and therefore be unbleachable. It may have been bleachable but mishandled in some processing step. Poor refining will leave residual phospholipid gums (lecithin) which will cause darkening of the oil during deodorization. Residual nickel catalyst in a hydrogenated oil will also cause darkening during deodorization. Vegetable oils and shortenings will darken on being stored for a long time, especially at elevated temperatures. Finally, a dark oil may be merely the result of the use of an insufficient amount of bleaching material.

Color of an oil is usually measured by the Lovibond Tintometer (AOCS Method Cc 13b-45). This is done by comparing the color of the oil in a tube of standard depth with the color of several glass standards. The colored glasses are graduated in three series: yellow, red, and blue. Most oils require only a combination of yellow and red glasses. The analyst selects the closest red glass, which is the easiest color to match, and selects a yellow glass with ten times the color value. He then may have to re-match the red and reselect the yellow glasses until he gets a perfect match.

The FAC color method (AOCS Method Cc 13b-45) uses a series of liquid color standards sealed in glass tubes to be compared with the oil sample in a tube of the same size. Solid glass standards are also available. The method is designed primarily for darker inedible oils.

Chlorophyll, the green pigment of plants, is also found in some oils. It is determined by absorption spectra, the amount of color absorbed by the oil sample at specific wavelengths of light. In this case, the wavelengths are 630, 670, and 710 nm. The results are expressed as ppm chlorophyll (AOCS Method Cc 13d-55). The Lovibond blue number has also been used, as yellow and blue combine to make green.

Shortening whiteness is measured by reflectance. Visual methods are not very accurate. Meters have been designed which shine a beam of light onto a smoothened surface of the sample and measure the amount of light reflected into a photosensitive tube.

Peanut butter and margarine colors can be graded visually by compari-son with color standards. These usually consist of colored plastic or enameled sticks arranged in gradations from light to dark. A color plate consisting of a series of photographs of potato chips increasing in color intensity from very light to very dark is available from the Potato Chip In-stitute International.

Reflectance meters are used for the determination of color of peanut butter, margarine, and mayonnaise by measuring the reflectance of light of different wavelengths and interpreting the values received in compari-son with those obtained from the measurement of selected standards (Mackinney and Little 1962).

BIBLIOGRAPHY

AMERICAN OIL CHEMISTS' SOC. 1967. Official Methods of Analysis. Am. Oil Chemists' Soc., Chicago.

BAILEY, A. E. 1950. Melting and Solidification of Fats. Interscience Pub-lishers Div., John Wiley & Sons, New York.

BLOOM, O. T. 1925. Consistency tester. U.S. Pat. 1,540,979. June 9.

BRAUN, W. Q. 1955. Dilatometric measurements. J. Am. Oil Chemists' Soc. 32, 633–637.

CLARDY, L., POHLE, W. D., and MEHLENBACHER, V. C. 1952. A shortening consistometer. J. Am. Oil Chemists' Soc. 29, 591–593.

DEUEL, H. J., JR. 1951. The Lipids—Their Chemistry and Biochemistry, Vol. I. Interscience Publishers Div., John Wiley & Sons, New York.

DuRoss, J. W., and Knightly, W. H. 1965. Relationship of sorbitan mono-stearate and polysorbate 60 to bloom resistance in properly tempered chocolate. Mfg. Confectioner 45, No. 7, 50–52, 54–56.

Ellis, B. H. 1967. Efficient use of sensory evaluation methods. Food Prod. Develop. 1, No. 5, 18, 19, 22, 32

Fulton, N. D., Lutton, E. S., and Wille, R. L. 1954. A quick dilatometric method for control and study of plastic fats. J. Am. Oil Chemists' Soc. 31, 98–103.

Gearhart, W. M., Stuckey, B. N., and Austin, J. J. 1957. Comparison of methods for testing the stability of fats and oils and of foods containing them. J. Am. Oil Chemists' Soc. 34, 427–430.

Hoerr, C. W., and Ziemba, J. V. 1965. Fat crystallography points the way to quality. Food Eng. 37, No. 5, 90–95.

Jacobson, G. A., Tiemstra, P. J., and Pohle, W. D. 1961. The differential cooling curve. A technique for measuring certain fat characteristics. J. Am. Oil Chemists' Soc. 38, 399–402.

Joffe, M. H. 1942. Mayonnaise and salad dressing products. Emulsol Co., Chicago.

Joyner, N. T., and McIntyre, J. T. 1938. The oven test as an index of keeping quality. Oil Soap 15, 184–186.

Mackinney, G., and Little, A. C. 1962. Color of Foods. Avi Publishing Co., Westport, Conn.

Mehlenbacher, V. C. 1960. The Analysis of Fats and Oils. Garrard Press, Champaign, Ill.

Merker, D. R., Brown, L. C., and Wiedermann, L. H. 1958. The relationship of polymorphism to the texture of margarine containing soybean and cottonseed oils. J. Am. Oil Chemists' Soc. 35, 130–133.

Norris, F. A., and Mattil, K. F. 1946. Interesterification reactions of triglycerides. Oil Soap 23, 289–291.

Pohle, W. D. 1969. Personal communication. Oak Brook, Ill.

Pohle, W. D. et al. 1964. A study of methods for evaluation of the stability of fats and shortenings. J. Am. Oil Chemists' Soc. 41, 795–798.

Quimby, O. T., Wille, R. L., and Lutton, E. S. 1953. On the glyceride composition of animal fats. J. Am. Oil Chemists' Soc. 30, 186–190.

Sherwin, E. R. 1968. Methods for stability and antioxidant measurement. J. Am. Oil Chemists' Soc. 45, 632A, 634A, 646A, 648A.

Smouse, T. H., and Chang, S. S. 1967. A systematic characterization of the reversion flavor of soybean oil. J. Am. Oil Chemists' Soc. 44, 509–514.

Swern, D. 1964. Bailey's Industrial Oil and Fat Products, 3rd Edition. Interscience Publishers Div., John Wiley & Sons, New York.

Weiss, T. J. 1963. Fats and oils. In Food Processing Operations, Vol. 2, M. A. Joslyn, and J. L. Heid (Editors). Avi Publishing Co., Westport, Conn.

Wiedermann, L. H. 1968. Margarine oil formulation and control. J. Am. Oil Chemists' Soc. 45, 515A, 520A–522A, 560A.

Wiedermann, L. H., Weiss, T. J., Jacobson, G. A., and Mattil, K. F. 1961. A comparison of sodium methoxide-treated lards, J. Am. Oil Chemists' Soc. 38, 389–395.

CHAPTER 2

Commercial Oil Sources

A large number of fats and oils are available for commercial use. Table 8 lists these oils with production and consumption figures for the United States for 1967. Fats and oils can be divided into categories according to their source. The vegetable oils represent the largest and most diverse grouping. They are also the most important from a commercial standpoint. Meat fats are used to a lesser but significant degree in food products. The marine oils are used as food fats in a number of countries but not in the United States. A group of synthetic fats, the acetoglycerides, have found a specialized utility in recent years.

TABLE 8

SUPPLY AND DISPOSITION OF FOOD FATS AND OILS FOR 1967

Source	Fats and Oils in Millions of Pounds		
	Production	Domestic Use	Export
Lard	2032	1802	242
Beef tallow	545	567	12
Cottonseed	1034	1090	52
Soybean	9004	5096	3964
Corn	452	432	25
Peanut	226	211	32

Source: U.S. Dept. Agr. (1969).

VEGETABLE FATS AND OILS

The vegetable oils are usually classified into groups according to their fatty acid composition. In this manner, the oils are divided into the lauric acid containing fats, e.g., coconut and the palm kernel oils; vegetable butters, e.g., cocoa butter; linoleic acid oils, e.g., cottonseed oil; linolenic acid oils, e.g., soybean oil; and erucic acid oils, e.g., rapeseed oil. In this treatise, however, the oils will be discussed roughly in order of their economic importance.

Soybean oil is by far the most valuable commercial oil. Cottonseed oil is second in importance. There then follows a group of oils which are useful as salad or cooking oils. Finally, a number of miscellaneous oils, including the laurics, will be discussed. The individual oils in this last group are each important for the specific fatty acids which they contain.

26

Soybean Oil

It is difficult to realize that soybean oil was hardly known in the United States prior to World War II. Since that time, the use of soybean oil has expanded in this country until its consumption for edible purposes amounts to 73% of all vegetable oils consumed. Its cost is usually somewhat below that of other vegetable oils.

Soybean oil has the following fatty acid composition and analytical constants:[1]

	%
Myristic acid	0.1
Palmitic acid	10.5
Stearic acid	3.2
Oleic acid	22.3
Linoleic acid	54.5
Linolenic acid	8.3
Arachidic acid	0.2
Eicosenoic acid	0.9
Iodine value	120–141
Melting point	$-23°$ to $-20°C$
Saponification value	189–195

Crude soybean oil contains about 1.8% phosphatides. It has become the major source of commercial lecithin. The sitosterols in soybean oil are partially removed by refining and have been used as a means of depressing blood cholesterol level.

Unhydrogenated soybean oil tends to revert to give flavors described as grassy, painty, or fishy. The oil is used commercially in the production of mayonnaise and salad dressings of all types. It is rarely used in frying as heat causes the development of fishy odors in the air which are quite objectionable. Freshly fried foods do not seem to retain fishy odors or flavors, but storage of foods fried in soybean oil will cause the foods to develop such off-odors and -flavors in a short time.

The poor stability of unhydrogenated soybean oil has been offset by hydrogenating it slightly to match cottonseed oil in iodine value. As with cottonseed oil, this hydrogenated soy oil will deposit crystalline fats on

[1] These and subsequent values are from Bull. *1170, Composition and Constants, Natural Fats and Oils*, Ashland Chemical Co. They represent GLC analyses of typical samples. Since oils are natural products, the normal oils will cover a range of values depending on plant variety, geographical source, climatic and seasonal variations. In the meat fats, variations will depend on the feedstuffs used in finishing the animals prior to slaughter.

being chilled. The oil is therefore winterized and sold in competition with cottonseed salad oil.

Partially hydrogenated soybean oil is the major component of vegetable shortenings and ordinary margarines as currently formulated. Fully hydrogenated soybean oil crystallizes in the beta phase which limits its usage in this form.

Cottonseed Oil

At one time, cottonseed oil was the major vegetable oil in the United States. It has surrendered this position to soybean oil for the reason that cottonseed oil production is dependent on the market for cotton fiber. The increased use of synthetic fibers has markedly decreased the need for cotton fiber and thereby the availability of seed for oil production.

Soybeans are an excellent source of high quality protein as well as of oil. The need for such protein is also great and is doubtlessly increasing. Cottonseed proteins have a limited usage thus far. This is due to the presence of gossypol, a pigment with slight to high toxicity depending on the animal species involved. Work is underway to eliminate or at least detoxify gossypol. A new group of cottonseed varieties is also being developed in which gossypol pigment glands are completely absent. Increasing the value of cottonseed protein could help reduce the fiber prices and improve the competitive position of cotton. This would increase the availability of cottonseed oil.

The following analysis has been obtained for cottonseed oil:

	%
Myristic acid	1.0
Palmitic acid	25.0
Palmitoleic acid	0.7
Stearic acid	2.8
Oleic acid	17.1
Linoleic acid	52.7
Iodine value	96.8–111.6
Melting point	$-2°$ to $2°C$
Saponification value	189–198

Unhydrogenated whole cottonseed oil (also known as summer oil or cooking oil) is used mostly for frying potato chips. The bulk of the available cottonseed oil is winterized to form salad oil. While some cottonseed salad oil is still used for mayonnaise and salad dressing manufacture, most finds its way into restaurant and household use as a general purpose oil.

The amount of actual stearine removed by winterization is approximately 10% of the whole oil. The total material removed may be as high as 30% due to entrainment of liquid oil in the solid stearine crystals. This depends on the overall process. As a result of this, the fatty acid composition of winter oil is not too different from that of summer oil. The palmitic acid content will be 3–4% lower and the unsaturates will be proportionately higher. The iodine value of the oil will increase by about five units.

Obviously the stearine obtained will be highly variable depending on the amount of entrained oil. Winterization from solvent will yield the hardest material. Soft stearines from nonsolvent winterization are sometimes recrystallized to increase the yield of oil. Oil is the desired fraction while the stearine is often a drug on the market. As much as 25% soft stearine may be added back to cottonseed summer oil for frying purposes without affecting performance of the oil.

The analysis of products obtained from the winterization of cottonseed oil is given in Table 9.

TABLE 9

COMPOSITION OF PRODUCTS RESULTING FROM THE
WINTERIZATION OF COTTONSEED OIL

Fatty Acid	Weight Percent of Fatty Acids in Oil Component			
	Cooking Oil (Unwinterized)	Salad Oil	Stearine from Oil	Stearine from Solvent
Myristic	0.8	0.7	0.6	0.4
Palmitic	24.2	22.6	32.4	52.0
Palmitoleic	0.6	0.4	0.3	0.0
Stearic	1.6	2.8	2.4	1.6
Oleic	21.0	19.8	17.2	12.0
Linoleic	51.8	53.7	47.1	34.0
Iodine value	107.8	111.4	98.0	71.1

Source: Weiss (1967).

Partial hydrogenation of cottonseed oil yields shortening bases which are not sufficiently different from partially hardened soybean oil to warrant the extra cost. Fully hydrogenated cottonseed oil, however, is stable in the beta-prime crystal form and imparts this property to shortenings prepared from it. Since no other vegetable oil available in sufficient quantity forms beta-prime crystals, cottonseed oil has a unique value to the shortening manufacturer.

Corn Oil

The corn oil of commerce is the oil obtained from the germ of the corn kernel. Corn gluten oil is also available in small amounts. This latter oil is found directly under the hull of the corn kernel. Corn (germ) oil is a relatively light colored oil with a characteristic winey odor. Corn gluten oil is dark red. It has a strong odor which on dilution is reminiscent of corn meal or pop corn.

The production of corn oil is related directly to the market demand for corn starch and corn starch products. As with cottonseed oil, the production of corn oil will always be limited.

The fatty acid composition and other analyses of corn oil are:

	%
Palmitic acid	11.5
Stearic acid	2.2
Oleic acid	26.6
Linoleic acid	58.7
Linolenic acid	0.8
Arachidic acid	0.2
Iodine value	124.4
Melting point	$-12°$ to $-10°C$
Saponification value	187–193

Corn oil is used primarily in unhydrogenated form. It contains waxes which should be removed by chilling if the oil is to remain clear under refrigeration. For many years, most corn oil was sold for household use. Some corn oil was also used for frying of chips and in salad dressings. The recent promotion of polyunsaturated oils for nutritional purposes has literally exploded the commercial usage of corn oil in various foods, especially in margarines containing liquid or unhydrogenated oils.

Hydrogenated corn oil is similar to hydrogenated soybean oil and offers no advantage over it. An occasional product will contain hydrogenated corn oil in order to carry the promotional claim "contains 100% corn oil."

Peanut Oil

Peanut oil comes from the cotyledons of the peanut. The germ contains a different oil but it is quantitatively negligible.

Peanuts are planted to be consumed as whole nuts or whole nut products such as peanut butter. The nuts are graded after harvest with mostly low grade nuts going into oil production. Peanut oil is, therefore, merely a by-product of a much larger industry.

Peanut oil has the following analysis:

	%
Palmitic acid	11.0
Stearic acid	2.3
Oleic acid	51.0
Linoleic acid	30.9
Arachidic acid	0.7
Behenic acid	2.3
Lignoceric acid	0.8
Iodine value	84–100
Melting point	−2°C
Saponification value	188–195

Unhydrogenated peanut oil can be deodorized to a bland oil which does not revert readily. When it does revert, the flavor is reminiscent of roasted peanuts. A number of individual commercial friers of various food and snack items have developed a preference for this oil.

On being chilled, peanut oil solidifies to the extent that it cannot be winterized. It cannot be used for manufacture of salad dressings.

At one time, fully hydrogenated peanut oil was considered to be the only proper stabilizer for use in peanut butter. It has since been supplanted to a great extent by other hard fats and monoglycerides, a fact which has been recognized in the newly promulgated U.S. FDA Standards of Identity for peanut butter.

Safflower Oil

Safflower oil is a latecomer to the edible oil market. It became popular with the demand for polyunsaturated oils. It has the highest linoleic acid content of any commercial oil. The safflower grows in semiarid land. It likes hot, dry air and moist soil. It thrives in areas and under conditions which are not always economical for growth of other crops. Unfortunately, safflower seed has a high fiber content which limits use of the oil meal as an ingredient in livestock feeding.

The composition and analysis of safflower oil is given as:

	%
Myristic acid	0.1
Palmitic acid	6.7
Stearic acid	2.7
Oleic acid	12.9
Linoleic acid	77.5

Arachidic acid	0.5
Eicosenoic acid	0.5
Iodine value	143.3
Melting point	$-18°$ to $-16°$C
Saponification value	190.1

A newly developed variety of safflower yields an oil in which the oleic and linoleic acid contents are reversed in magnitude. While this oil could not be considered as highly polyunsaturated, it could be used as a salad oil.

Regular safflower oil is used wherever high polyunsaturated fatty acid content in foodstuffs is desired. It is used in production of special mayonnaise, salad dressings, liquid oil margarines, dietetic mellorine, etc. As with soybean oil, the flavor of safflower oil is not stable in frying. For this reason, it has not been too successful as a household oil when used by itself. Some of it is still available however. Blends of safflower with cottonseed oil have proven to be acceptable since the cottonseed oil odor and flavor predominates on frying.

Obviously, hydrogenation destroys the polyunsaturated fatty acids and serves no purpose in the processing of safflower oil.

Sunflower Oil

The use of sunflower seed as a source of oil is still experimental in the United States. Sunflower oil is commercially important in Canada and Russia. It is one of the few vegetable oils that can be produced in cold climates.

Sunflower plantings in the United States, mostly in the Dakotas, have been primarily to produce seeds for feeding of birds. Southern planters have been searching for oilseed crops to replace the diminishing cotton crop. Sunflower has seemed to be one of the more promising of these.

Sunflower oil from cold climates has the following analysis:

	%
Palmitic acid	7.0
Stearic acid	3.3
Oleic acid	14.3
Linoleic acid	75.4
Iodine value range	125–136
Melting point	$-18°$ to $-16°$C
Saponification value	188–194

Sunflower oil is similar in stability to safflower oil. Oil grown in cold climate does not require winterization but some oils grown in the southern parts of the United States have been found to deposit stearine on being chilled. The southern oils have been found to contain over 50% oleic and as little as 35% linoleic acids with an iodine value about 106.

Sunflower oil has the same potential utility as safflower oil.

Olive Oil

Olive oil is the oil of the Mediterranean countries. In the United States, this oil is considered as a gourmet item. The oil is normally pressed from the fruit of the olive tree, but not from the kernel of the pit. This oil is known as *virgin* olive oil. It is never deodorized. It is highly flavored and highly prized by those who prefer it for this reason. Virgin olive oil sold in the United States may be imported or may be produced from olives grown in California.

The residues from pressing olive pulp for oil, lower grade olives, and olive pit kernels are extracted with solvent to remove the oil. The solvent is stripped from the oil which is finished for use by deodorization. It is bland in flavor. Most of the olive oil imported into the United States is of this type. It is usually labeled as *pure* olive oil, but cannot be called virgin oil.

Olive oil is very expensive in relation to other edible oils. For this reason, a large number of blends of olive and soybean or cottonseed oils is available on the market. The amount of olive oil in such blends is small, perhaps in the order of 5–10%. If the blend contains virgin olive oil, the flavor of this oil is sufficiently strong to please the palate of a public used to perfectly bland oils. Obviously, blends of olive and cottonseed oils can be used both for salad dressings and for frying, while olive-soybean oil blends should be restricted to salad dressing use.

Olive oil has the following analysis:

	%
Palmitic acid	16.9
Palmitoleic acid	1.8
Stearic acid	2.7
Oleic acid	61.9
Linoleic acid	14.8
Linolenic acid	0.6
Arachidic acid	0.4
Eicosanoic acid	0.1
Behenic acid	0.2
Iodine value	80–88
Saponification value	188–196

Rapeseed Oil

Like the sunflower, the plants from which rapeseed is gathered are capable of growing in cold climates. Rapeseed oil, also known as *colza* oil when sold for salad oil uses, is popular in France and is a major source of margarine and shortening oils in Poland. The Canadian government has been promoting the planting of rapeseed as a domestic source of vegetable oil. Rape is planted in this country primarily for livestock forage, but oil millers are considering it as a potential oilseed crop as well. This is especially true for the cotton producing areas.

Wild mustard often grows as a weed in rapeseed plantings. The seeds of both plants are similar to each other as are the oils obtained from them. They are not always separated before being processed into oil. Both plants are in the cabbage family and contain sulfur and nitrogen bearing compounds. These give the crude oil a strong cabbage-like odor and flavor.

At one time, it was not unusual to find production lots of rapeseed oil which could not be hydrogenated. Residual sulfur compounds were apparently poisoning the hydrogenation catalyst. It was found that water-washing or degumming the crude oil prior to alkali refining and bleaching would effectively remove the interfering compounds. Many oil millers offer only degummed rapeseed oil to the market in order to avoid subsequent mishandling by the oil refiner.

Rapeseed oil has a very high level of fatty acids with chain lengths of 20 to 22 carbon atoms.

The composition and other analysis of rapeseed oil is given as:

	%
Myristic acid	0.1
Palmitic acid	4.0
Palmitoleic acid	0.1
Stearic acid	1.3
Oleic acid	17.4
Linoleic acid	12.7
Linolenic acid	5.3
Arachidic acid	0.9
Eicosenoic acid	10.4
Behenic acid	0.7
Erucic acid	45.6
Docosadienoic acid	0.1
Lignoceric acid	0.2
Tetracosenoic acid	0.6

Iodine value	81.4
Melting point	−9°C
Saponification value	170–180

Rapeseed oil would have nothing unusual to offer if it were to be used only as a salad oil. As a fully hydrogenated hardfat, however, its high content of long chain-length fatty acids makes this oil unique. The long chain hardfats solidify into an unusually small beta-prime crystal with special properties for production of fluid margarine and stabilization of peanut butter. Some work on the effectiveness of distilled arachidic and behenic acid monoglycerides for cake baking (Handschumaker and Hoyer 1961) suggests that further investigation of hydrogenated rapeseed oil monoglycerides for various uses is in order.

At this writing, the investigation of crambe oil is underway. This oil contains over 50% very long chained fatty acids. It would seem that fully hydrogenated crambe oil should have some very interesting functional properties.

On the other extreme, a variety of rapeseed has been developed which has practically no 20 to 22 carbon-chain fatty acids. This may be satisfactory for salad oil use but such oil would not have the unique properties of conventional rapeseed oil when fully hydrogenated.

Coconut Oil

The lauric acid fats have unusual properties when compared to other edible fats and oils. While they are fairly saturated, the presence of the short chained acids at levels of about 50% makes these oils solid at slightly below room temperature and liquid at slightly above that point.

Coconut oil is the most abundant of the lauric acid oils. It comes from the kernel of the coconut palm as opposed to the other lauric oils which come from the kernel of other types of palm.

The fatty acid composition and analysis of coconut oil is:

	%
Caprylic acid	7.6
Capric acid	7.3
Lauric acid	48.2
Myristic acid	16.6
Palmitic acid	8.0
Palmitoleic acid	1.0
Stearic acid	3.8
Oleic acid	5.0
Linoleic acid	2.5

Iodine value	7.5–10.5
Melting point	23°–26°C
Saponification value	250–264

Coconut oil is sold mostly as 76° and 92° oils. The numbers refer to the Wiley melting point of the products. The 76° oil is unhydrogenated while the 92° oil has been partially hydrogenated. A fully hardened (110°F) oil is also available.

The unhydrogenated oil is used for preparation of coatings for ice cream bars since it is very hard on the bar but melts quickly and completely in the mouth. It has a high degree of stability against oxidation due to its low level of unsaturated fatty acids. This makes it useful for the frying of nuts and snacks where a long shelf-life is desired. The short chained acid fats also have a lower viscosity than ordinary oils. They therefore feel less greasy in the mouth. This makes them desirable as spray coatings for cereals and crackers and as lubricants in various confections such as caramels and nougats.

Hydrogenated coconut oil is often used in imitation dairy products in place of butterfat. It is also used in the preparation of confectioners' hard butters.

Coconut oil also has its limitations. Hydrolysis of the oil, even to a small extent, liberates the short chained fatty acids which are highly flavored. So-called hydrolytic rancidity causes the development of a disagreeable soapy flavor in coconut oil. This happens slowly in the presence of free moisture alone, rapidly where the enzyme lipase is present. Unblanched nuts and low-temperature pasteurized milk products contain active lipase and should not be used with coconut oil. The shelf-life of imitation fluid dairy products made with coconut oil is limited even in the absence of lipase.

Coconut oil is an excellent frying oil when used alone. For some as yet unknown reason, blends of coconut with nonlauric oils foam badly when used for deep fat frying. This same foaming tendency could prevent spattering in pan frying.

Palm Kernel Oil

Palm kernel oil is a composite of a number of specific oils, such as babassu, tucum, gru-gru, ouricuri, and the like. The exact composition will vary with whatever source is available at any given time. Some of the palms are plantation grown and of specific variety, others are wild jungle palms and of mixed variety. The kernels are hard white nuts and are not at all like the large, hollow coconut. The oil is more unsaturated than coconut oil as can be seen from the following analysis:

	%
Caprylic acid	1.4
Capric acid	2.9
Lauric acid	50.9
Myristic acid	18.4
Palmitic acid	8.7
Stearic acid	1.9
Oleic acid	14.6
Linoleic acid	1.2
Iodine value	14–23
Melting point	24°–26°C
Saponification value	245–255

Palm kernel oil is not as available as and is more expensive than coconut oil. It is more solid than coconut oil and is fractionated to yield a hard, sharply melting product useful in certain types of confectionery coatings. Palm kernel oil is capable of being hydrogenated further than coconut oil since it is more unsaturated. This makes it useful for the preparation of other specialized confectioners' hard butters which have characteristics somewhat different from those obtained from coconut oil.

Palm Oil

The husk surrounding the cluster of palm kernels yields an oil, or more correctly, a fat known as palm oil. It is entirely different from palm kernel oil and contains no lauric acid.

Crude palm oil has a deep orange color due to the presence of carotene pigments. It was once used to color margarine. The oil also has a high free fatty acid content which makes it difficult to process. The FFA can be 1–50% but usually runs 5–15%.

Palm oil has the following analysis:

	%
Lauric acid	0.1
Myristic acid	1.2
Palmitic acid	46.8
Stearic acid	3.8
Oleic acid	37.6
Linoleic acid	10.0
Arachidic acid	0.2
Eicosenoic acid	0.3

Iodine value   48–56
Melting point   27°–50°C
Saponification value   196–202

Palm oil is not used to any great extent in the United States. If the price is low enough to be competitive it is often "lost" in shortenings. It is similar to lard in hardness but is stable in the beta-prime crystalline phase. This would make fully hydrogenated palm oil behave more like cottonseed oil or its stearine when hardened to the same iodine value.

Cocoa Butter

Vegetable butters are fats of vegetable origin that are hard at room temperature but melt quickly at slightly elevated temperatures. Vegetable tallows, e.g., Borneo tallow, are similar but have much higher melting points. Cocoa butter is the major commercial vegetable butter, although others, e,g., shea butter, are available in limited quantities. These fats have been used occasionally as cocoa butter substitutes. They are more commonly used in Europe.

Cocoa butter is unique among the vegetable butters in possessing a delightful, characteristic chocolate odor. Contrary to popular opinion, cocoa butter has practically no flavor. Occasional residues of cocoa powder in cocoa butter contribute a bitterness. Cocoa powder is easily

TABLE 10

SOLID FAT COMPOSITION OF COCOA BUTTER IN RELATION TO TEMPERATURE

Temperature, °C	% Solid Fats
0	99.4
5	98.1
10	95.9
15	93.2
20	89.2
25	83.3
30	63.9
34.1	0.0

Source: Landmann *et al.* (1960).

detected when present as it is insoluble in the butter. It appears as a dark cloud when the butter is melted. The production of cocoa butter and related chocolate products are discussed in Chap. 12.

Cocoa butter melts sharply because the major triglyceride components are of a single type, primarily 1,3-disaturated, 2-monounsaturated triglycerides (Chacko and Perkins 1964; Scholfield and Dutton 1959). Cocoa butter is perhaps the purest natural fat in this respect. A mixture of disaturated triglycerides with unsaturated fatty acids in both the 1- and

2-positions would melt less sharply. The sharpness of melting of cocoa butter expressed in terms of the change in percent solid fats with temperature is given in Table 10. These values should be compared with those for a normal shortening in Table 4.

Cocoa butter has the following analysis:

	%
Myristic acid	0.5
Palmitic acid	25.0
Stearic acid	34.5
Oleic acid	36.5
Linoleic acid	3.0
Linolenic acid	0.5
Iodine value	35–40
Melting point	35°C
Saponification value	194

The edible uses for cocoa butter are restricted to the production of confections and related goods. The high price of cocoa butter in relation to that of other fats has led to the large scale development of various substitutes for cocoa butter. An entire segment of the fat and oil industry is devoted to this specialized field.

MEAT FATS

There are only two importatn fats of (land) animal origin in commercial usage, lard and tallow. There are a number of varieties of these fats based on production methods and type of raw fat stock.

At the beginning of this century, lard was the major shortening product. Lard is a naturally hard fat with a pleasant, natural flavor. It did not require hydrogenation or deodorization to be useful. These processes were developed in order to produce shortenings from the vegetable oils.

Eventually, vegetable oils supplanted lard as the major source of shortening fats. Hydrogenated cottonseed oil shortenings were found to cream better than lard shortenings. This led to the development of new baked goods which relied on improved creamability of the shortening. Further work led to the molecular rearrangement of lard. This interesterification process yielded a lard product equal in creaming quality to cottonseed oil shortenings.

In many but not all areas, it is now possible to use meat fats or hydrogenated vegetable oil shortenings interchangeably. When performance is the only criterion, the user can save money by purchasing the lowest cost product, most frequently the one made from meat fats. Often, however, prejudices rule out one type of product, again most frequently, the one made from meat fats.

Lard

Lard is defined as the fat rendered from the fatty tissues of the pig. There are more specific definitions established under the U.S. Dept. of Agr. Bureau of Animal Industry. For example, there are numerous fatty tissues. Lard must come primarily from visceral depot fats and subcutaneous fatty tissues. If skin fats, bone fats, cured ham facings, and the like are used the product must be called "rendered pork fat."

Depot fats are relatively hard, especially those surrounding the kidney. They have been termed "killing" fats since they are removed during the slaughtering operations. The fatty layers found in the back and belly muscles and other subcutaneous areas yield a softer fat. They are the "cutting" fats obtained when the hog carcass is cut into its various parts. Average lard consists of about 65% cutting and 35% killing fats (Mattil 1964).

Normal lard has the following analysis:

	%
Myristic acid	1.5
Palmitic acid	27.0
Palmitoleic acid	3.0
Margaric acid	0.5
Stearic acid	13.5
Oleic acid	43.5
Linoleic acid	10.5
Linolenic acid	0.5
Iodine value	58–68
Melting point	42°C
Saponification value	195

The iodine value of killing fats is 57–60, of cutting fats, 67–70, and of an ordinary blend of the two, 62–65. A small amount of lard is available occasionally from hogs finished on peanuts by allowing the animals to root in peanut fields following harvest. The iodine value of such lard is usually over 85.

The SFI curve of lard has a characteristic shape not found in conjunction with any other fat. Figure 10 shows this curve in comparison with one for beef tallow. The overall shape of the tallow curve, however, is common to many hydrogenated vegetable shortenings.

Lard normally crystallizes in the beta phase. Beta crystalline short-enings give very poor results in cake baking and cake icing preparation. On the other hand, beta crystals are preferred for the baking of pie crusts as they seem to contribute a flakiness to the crust.

Interesterification modifies the lard crystal structure to a beta-prime type. The SFI curve is also changed as is shown in Fig. 10. Hydrogena-

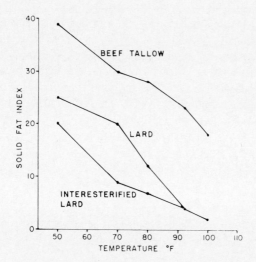

FIG. 10. SFI VALUES OF THE MEAT FATS

tion increases the stability of lard. Thus, lard is one of the most versatile of fat sources.

Unhydrogenated and partially hydrogenated lards are used for commercial frying. Partially hydrogenated lard is also used as a general shortening base.

Lard is fractionated by pressing the chilled fat to yield about 60% lard oil as the primary product and lard stearine as a by-product. Lard oil had found great favor as a bread pan lubricant probably because of the natural lard flavor it possessed. Lard stearine was disposed of in producing firm lard for pie crusts. The production of lard oil by pressing involves a considerable amount of hand labor and is, therefore, a costly process. It is no longer practiced on a commercial basis in large scale.

Fully hydrogenated lard is used to firm up unhydrogenated lard for packaging purposes. It is also added wherever a firm lard is desired for specific applications.

Lard has the following composition:

	%
Lauric acid	0.3
Myristic acid	1.7
Myristoleic acid	0.2
Pentadecanoic acid	0.1
Palmitic acid	26.2
Palmitoleic acid	4.0
Margaric acid	0.5
Heptadecenoic acid	0.3
Stearic acid	13.5
Oleic acid	42.9
Linoleic acid	9.0
Linolenic acid	0.3
Arachidic acid	0.2
Eicosenoic acid	0.8
Iodine value	53–77
Melting point	33°–46°C
Saponification value	190–202

Tallow

As with lard, the term "tallow" covers a broad spectrum of actual products. Tallow is normally thought of as beef fat. It may contain some mutton tallow or lard.

Beef fat, like lard, occurs as cutting and killing fats. Unlike pork processing, where slaughter of the hog and butchering of the carcass into smaller cuts often takes place under the same roof, the beef carcass may be shipped to market as sides to be cut further at a location far from the slaughtering site. The trimmings of fatty tissue (cutting fats) are then collected for rendering and returned to a nearby processing plant. It is not unusual to find that pork and lamb trimmings have been mixed with the beef trimmings.

Tallow is usually dry rendered. The fat is then deodorized at some point in processing of the finished shortening. Tallow has a strong meaty flavor.

A carefully processed wet rendered beef fat was once a major source of edible fat. The product known as oleo stock, had a rich buttery flavor. It was separated by pressing into oleo oil and oleo stearine. Unlike tallow, the mild oleo flavor was highly desirable and was never removed by deodorization. Oleo oil was once prepared into margarine by emulsification with milk. More recently it was used by candy manufacturers as a replacement for butter oil since it had a similar flavor and similar melting characteristics. Oleo stearine was a hard, waxy fat which was used for manufacture of puff paste. Oleo stock and its fractions are no longer commerically available in great quantity.

Tallow has the following composition:

	%
Myristic acid	3.1
Myristoleic acid	0.4
Palmitic acid	29.1
Palmitoleic acid	3.4
Margaric acid	0.4
Heptadecenoic acid	0.4
Stearic acid	18.9
Oleic acid	44.0
Linoleic acid	0.3
Iodine value	45–55
Melting point	48°C
Saponification value	190–199

The SFI curve for tallow is given in Fig. 10. Tallow crystallizes in the beta-prime phase. It is a naturally hard material and is frequently used without hydrogenation in shortening and bakery margarine formulations. Where increased stability is required, tallow is often blended with soybean oil before hydrogenation in order to prevent it from becoming too hard to be practical.

Fully hydrogenated tallow is used in meat fat shortenings in the same manner as is fully hydrogenated cottonseed oil in vegetable shortenings. They both have beta-prime crystal structure.

MARINE OILS

The major marine oils are from the herring, menhaden, pilchard, sardine, and whale. The last of these is commercially the most important. While the whale is a warm blooded mammal, its body fat is similar to that of other marine animals.

Marine oils are not used for edible purposes in the United States. However, they are used extensively in Canada and Europe for the manufacture of shortenings and margarine.

Marine oils are highly unsaturated. The unhydrogenated oils revert and oxidize quickly to develop extremely disagreeable fishy flavors and odors. The oils become fairly stable on hydrogenation. The hydrogenated marine oils revert in flavor to develop a characteristic sweet odor and flavor somewhat reminiscent of undeodorized fully hydrogenated vegetable oil. Shortening manufacturers in the United States have been reluctant to work with marine oils because of these flavor problems. This situation will probably remain unchanged as long as there is an abundance of meat fats and vegetable oils in this country.

Fatty acid composition of the marine oils is given in Table 11. The large variety of unsaturated fatty acids is characteristic of all marine oils.

TABLE 11

COMPOSITION OF MARINE OILS

Fatty Acid		Oil Source				
No. C. Atoms	No. Double Bonds	Herring	Menhaden	Pilchard	Sardine	Whale
14	0	1.3	7.3	5	6	8
14	1	...	...	...	...	2
15	0	0.3	0.4	...	...	...
16	0	13.6	23.6	15	10	12
16	1	10.2	9.9	12	13	15
17	0	0.5	0.9	...	...	...
18	0	1.4	2.6	3	2	2
18	1	23.5	17.0	6	24	33
18	2	0.8	1.2	12	...	9
18	4	11.2	4.1	...	...	...
19	0	2.0	1.2	...	...	...
20	2	3.0	0.3	...	...	...
20	3	...	0.2	...	...	...
20	4	9.1	3.4	18	26	8
20	5	13.1	12.0	...	...	...
22	5	0.6	1.7	14	19	11
22	6	5.6	9.1	...	...	...
24	1	0.4	0.8	...	...	...
24	6	...	...	15	...	...
Iodine value		148	178	184–200	170–193	110–135

Source: Bull. *1170, Composition and Constants, Natural Fats and Oils*, Ashland Chemical Co., Div. of Ashland Oil and Refining Co.

SYNTHETIC FATS

The synthetic fats referred to here are the various acetoglycerides. They do not occur in nature. Acetoglycerides contain acetic acid as well as fatty acids. In one sense acetic acid may be considered as a fatty acid with a two carbon chain length. On the other hand, the acetyl radical is partially soluble in water. It does not behave in the same manner as the true fatty acids.

There are two basic methods for the preparation of acetoglycerides (Feuge 1955). One of these calls for the interesterification of any triglyceride with triacetin (glycerin triacetate). The acetic acid radicals are distributed randomly with the other fatty acid radicals on the glycerin molecule. The other method of preparation involves the reaction of distilled monoglycerides with acetic anhydride.

Acetoglycerides produced by this second reaction have been approved for food use by the U.S. FDA. Several products are available on the market. These were discussed by Brokaw (1960). They include reaction products of a fully hydrogenated monoglyceride where 50 and 70% of the available hydroxyl groups have been acetylated. Both are waxy solids, stable in the alpha crystalline form. They are also quite flexible.

Acetylated monoglycerides are useful as coatings for various foods (Luce 1967). The product containing the higher level of acetylation has a 99°–104°F melting point. It is useful as a protective moisture-proof coating for refrigerated foods. The acetoglyceride with 50% acetylation is potentially more versatile. It melts at 106°–109°F. It can serve as a protective coating for nonrefrigerated foods (Shea 1965). It also has emulsifying properties (Gehrke et al. 1959) making it useful in cake baking (Baur and Lange 1952) and for whipped toppings (Anon. 1968).

A third acetoglyceride consists of a distilled monoglyceride prepared from partially hardened fat (iodine value 40) reacted to acetylate 90% of the free hydroxyl groups. The product is an oil with 45°F congeal point. It is quite resistant to oxidation, making it useful as a lubricant for food machinery, a slab grease, and pan release agent. It is also recommended as a coating for dried fruits and nuts, cheese and shell eggs.

BIBLIOGRAPHY

Anon. 1968. Acetylated monoglycerides make dry base toppings taste and look like whipped cream. Food Process. 29, No. 5, 26.

Baur, F. J., and Lange, W. 1952. Plastic shortenings and process of producing same. U.S. Pat. 2,614,937. Oct. 21.

Brokaw, G. Y. 1960. FDA approval of distilled acetylated monoglycerides gives food processors new packaging techniques, edible protective coating. New ingredient, edible emulsifier, new processing aid, edible lubricant. Food Process. 21, No. 12, 38–40, 43, 46.

Chacko, G. K., and Perkins, E. G. 1964. Glyceride structure of cocoa butter. J. Am. Oil Chemists' Soc. 41, 843.

Eckey, E. W. 1954. Vegetable Fats and Oils. Reinhold Publishing Corp., New York.

Feuge, R. O. 1955. Acetoglycerides—new fat products of potential value to the food industry. Food Technol. 9, 314–318.

Gehrke, A. F., Marmor, R. A., Henry, W. F., and Greenberg, S. I. 1959. Cake shortening containing acetylated monoglycerides and method of making same. U.S. Pat. 2,882,167. Apr. 14.

Handschumaker, E., and Hoyer, H. G. 1961. Fluid shortening and method of making the same. U.S. Pat. 2,999,755. Sept. 12.

Landmann, W., Feuge, R. O., and Lovegren, N. V. 1960. Melting and dilatometric behavior of 2-oleopalmitostearin and 2-oleodistearin. J. Am. Oil Chemists' Soc. 37, 638–643.

LUCE, G. T. 1967. Acetylated monoglycerides as coatings for selected foods. Food Technol. *21*, 1462–1463, 1466, 1468.

MATTIL, K. F. 1964. Plastic shortening agents. *In* Bailey's Industrial Oil and Fat Products, 3rd Edition, D. Swern (Editor). Interscience Publishers Div., John Wiley & Sons, New York.

SCHOLFIELD, C. R., and DUTTON, H. J. 1959. Glyceride structure of vegetable oils by countercurrent distribution. IV. Cocoa butter. J. Am. Oil Chemists' Soc. *36*, 325–328.

SHEA, R. 1965. New edible protective coating keeps nuts fresh. Food Process. *26*, No. 5, 148–150.

U.S. Dept. of Agr. 1969. Fats and oils situation. Bull. *F.O.S.-247.* Washington, D.C.

WEISS, T. J. 1967. Salad oil manufacture and control. J. Am. Oil Chemists' Soc. *44*, 146A, 148A, 186A, 197A.

CHAPTER 3

Basic Processing of Fats and Oils

INTRODUCTION

Oil processing methods will not be given in great detail in this book. Other texts are available which cover this subject matter quite thoroughly. The broader aspects of fat and oil processing will be discussed in order to enhance general understanding of the field.

Certain processes are used for preparation of all fatty materials. These will be discussed in the sequence normally followed in refinery operations. Other specialized procedures which are unrelated to each other will be presented individually but in no particular order.

In most cases, equipment in use today is made from stainless steel. Older equipment was fabricated from black iron. Traces of rust would darken the first batches of oil exposed to freshly cleaned black iron. This was especially true when the shortening formulation contained monoglycerides. The metal surface soon became coated with a film of polymerized oil which protected subsequent batches of oil from iron contamination.

Copper and copper alloys must never be allowed to come in contact with oil products. Copper is an extremely active prooxidant which will cause rapid deterioration of fats and oils when present at as low a level as 0.01 ppm.

OIL EXTRACTION

The removal of fats and oils from their natural sources is the first step in the overall process. Each type of oil source requires special techniques which are usually not applicable to other sources. Rendering, pressing, and solvent extraction are the most common methods for the recovery of fats and oils.

Rendering

Rendering is one of the oldest ways of obtaining fats in free form. It consists of heating fatty tissues trimmed from animal carcasses after slaughter. The cooking process breaks down the cell wall of the tissue to release the fat (Bates 1968).

Lard is rendered by a number of different techniques. The products obtained differ primarily in flavor.

Open kettle rendering, as the name implies, involves cooking raw lard tissues to dryness in an open vessel. The lard has a strong but pleasant meaty flavor. This cooking method is rarely used today although it was once the preferred process.

Prime steam lard is prepared by cooking the tissues with steam at 30–60 psi in a closed vessel until the proteinaceous matter in the cell wall is denatured. The lard is allowed to float free from the tissues. It is then drawn off after standing for a sufficient length of time to form a separate layer. Prime steam lard has a mild but definite meaty flavor.

Dry rendered lard is produced by cooking the fatty tissues to dryness in a closed vessel under either atmospheric pressure or vacuum. The flavor of this product is harsher than that of prime steam lard. The latter is often sold in undeodorized form for use in bread and pastry baking. Dry rendered lard is deodorized and stabilized for sale as such or incorporated into shortening compositions which are finished by deodorization.

Low-temperature rendering is the most recent development in lard production (Downing 1959). In this process, the tissue cells are ruptured mechanically by comminuting at 115°–120°F, a temperature just sufficient to melt the lard but not high enough to denature the protein. The meaty tissues are removed by centrifugation. The lard is then clarified by passing it through a second centrifuge. The entire operation is continuous. Lard produced in this manner is mild, almost devoid of flavor. It also has a lower free fatty acid content and a greater stability against oxidation than lard produced by other methods.

Edible beef tallow is usually produced by the dry rendering process. A special low-temperature rendering was once applied to beef fats to yield a product called oleo stock. This material had a pleasant, buttery flavor. It was fractionated by pressing the partially crystallized stock into oleo stearine and oleo oil.

Pressing

Cold Pressing.—Another process for oil recovery which goes back into antiquity is cold pressing. High oil content seeds, such as sesame and peanut, and the oily pulp of olives yield free oil by the simple application of pressure. Oils of this type required no further processing. Sesame and peanut oils have a pleasant nutty flavor. Olive oil has a strong but well accepted flavor. However, cold pressing is not very efficient.

Hot Pressing.—The oil meals (seed residues from which oil has been removed) remaining from cold pressing contained an excessive amount of valuable oil. This led to the development of more efficient presses, such as the hydraulic batch press and the continuous screw press or expeller. These presses develop pressures of 1–15 tons per square inch, leaving 2–4% oil in the meal.

Unfortunately, such presses also developed excessive heat. This caused darkening of the oil and denaturation of oilseed protein. The prepress-solvent method was introduced to reduce damage to oil seeds during crushing. Here the seed is pressed to remove only part of the available oil. The temperature at this point is not sufficient to degrade the oil or protein to any great degree. Solvent extraction is then used to remove the balance of oil from the meal.

The prepress-solvent procedure was really a compromise measure. It would have been economically impractical for existing oil mills to discard their presses and install solvent systems to extract the entire amount of oil from the seed. Solvent extraction and solvent recovery equipment can be relatively small and therefore a less expensive installation where only a small part of the total available oil is removed by solvent.

Solvent Extraction

Extraction with solvent will recover practically all of the oil from seeds. Solvent oil meal contains about 0.5% residual oil. Extraction is also required with many seeds which have an insufficient amount of oil to make pressing feasible.

In solvent extraction of seeds, such as soybeans, the beans are first flaked and toasted or cooked. Flaking exposes as much surface of the bean to the solvent as is possible. Cooking denatures cell tissues so that the solvent can penetrate it more readily. The solvent, usually hexane, is then allowed to flow through the bean flakes to extract the oil. The solution of oil in solvent is referred to as the miscella.

At this point, the solvent is usually stripped off by evaporation to yield a crude oil for further processing. Current research is aimed at developing methods for processing oil in the miscella and removing the solvent only after the oil is in a finished state. Some plants are in operation at this time in which refining and winterizing steps are carried out on cottonseed oil in solvent directly following extraction of the oil.

Fats and oils which have desirable natural flavors, such as olive oil, cannot be extracted with solvent. Removal of solvent necessitates concurrent removal of flavor components. Meat fats do not require solvent extraction as the fatty tissues are easily processed by rendering methods.

DEGUMMING

Crude oils contain a large number of compounds of fatty nature other than triglycerides. The types and amounts vary with the oils involved. Production of high-quality finished oils requires the removal of as much of the nontriglyceride materials as possible. An exception to this is in the beneficial effect of residual tocopherols which act as antioxidants in vegetable oils.

Some of the extraneous material is water hydratable. Washing the oil with water or degumming makes these compounds insoluble in oil so that they may be removed by settling, filtration, or centrifugation. The crude material as removed is referred to as gums. Gums become crude lecithin on being dehydrated. They are a mixture of phospholipids and glycolipids.

As was previously discussed, degumming of soybean oil is practiced on a large scale due to the demand for soy lecithin. Corn oil lecithin is also marketed, but to a lesser extent.

The gums are not all removed with a single washing. Double and triple degummed soybean oils are available, each containing a lesser amount of phosphorus-bearing materials. Even with this, there is still some residual unhydratable phospholipids in degummed soybean oil.

REFINING

Refining usually refers to the removal of nonglyceride fatty materials by washing the oils with strongly alkaline water solutions. In addition to the gums removed by water-washing, the caustic materials, usually sodium hydroxide or sodium carbonate, also remove acidic compounds such as free fatty acids. The type of phospholipid residual in degummed oils is readily removed by caustic refining, provided the concentration of caustic in water solution is sufficiently high, in the order of 10–15% sodium hydroxide.

The physical process is to intimately mix crude oil and caustic solution and to separate the clear oil from the semisolid mass of impurities. The amount of caustic used is based on the free fatty acid content. These acids are neutralized to form soaps which are insoluble in the free oil. Excessive caustic can saponify the neutral oil, increasing the amount of fatty acids removed and reducing the yield of refined oil.

All refining was once done in a jacketed tank or kettle. The crude oil-caustic mix was agitated in an open kettle, heated and allowed to stand so that the soaps could settle. The settled material was called the "foots" or "soapstock." The bulk of this residue consisted of crude soaps to be sold to the soap maker for further processing. Much of this material is now neutralized with sulfuric acid to form a mixture of crude free fatty acids and phospholipids. It is sold to animal feed manufacturers as "acidulated foots."

Although some kettle refining is still being practiced to process small batches of special oils, most oils are now refined by continuous methods. Here the hot oil stream is injected by a proportioning system with a definite ratio of caustic to oil. The mixture is intimately mixed, heated, and separated by centrifugation. The process is not only faster but centrifugal separation of foots from oil is more efficient. These foots contain less entrained oil and result in a higher neutral oil yield.

After refining, it is necessary to remove residual soaps which were not completely eliminated in the first centrifugation. This is done by a water washing step. In this, refined oil is thoroughly mixed with hot water and passed through another bank of centrifuges. The washed oil is then dried under vacuum.

Attempts have been made to use degummed soybean oil as a substitute for refined oil in all uses. The residual phospholipids in degummed oil, however, have interfered with further processing. They have slowed filtration following bleaching. They have partially poisoned hydrogenation catalysts. They have caused darkening in deodorization and poor flavor stability in finished unhydrogenated oil. Alkali refining of degummed oil corrects this but reduces the economic advantage of using degummed oil. In fact, entrainment losses in refining added to similar losses in degumming may make degummed oil a rather poor bargain. A solution to this problem seems to be in the blending of degummed with refined oils. A good grade of degummed oil may be added to refined oil at up to a 25% degummed oil level without seriously affecting performance in various subsequent processes.

Steam refining is practiced on some fats and oils. In reality, this is a form of deodorization which will be discussed as such later. Steam refining is the stripping of free fatty acids from the oil by steam distillation under high vacuum and high temperature. It is satisfactory for processing meat fats which have little more than free fatty acids to be removed by refining. Safflower oil is often steam refined as, in this case, the quality of such oil is not much less than the quality of alkali refined oil. In addition, safflower oil is frequently blended with other oils or prepared into products which mask the relatively poor stability of this oil.

As was brought out in the discussion of degummed versus refined soybean oil, steam refining of soybean oil results in an inferior product and is not to be recommended.

BLEACHING

Bleaching is the process used for the removal of pigments from the oil although some color bodies are removed during other processes as well. Alkali refining removes water-soluble and acidic pigments. Deodorization removes volatile or at least steam distillable pigments and heat-sensitive pigments which decompose to yield colorless compounds. Hydrogenation also destroys certain reducible pigments.

The process of bleaching is relatively simple. Bleaching is carried out by mixing an adsorbent material with the oil to be treated, heating the oil to activate the adsorbent (by driving off moisture) and filtering the oil to remove the adsorbent along with the pigments to be removed. Atmospheric bleaching in open tanks and filtering in unenclosed plate and frame presses is still being practiced although vacuum bleaching is superseding it.

Three major types of adsorbents are used: neutral bleaching clay, acid processed bleaching clay, and activated charcoal. Bleaching clay is also known as fuller's earth. Acid earths are made from certain clays by treatment with sulfuric or sometimes with hydrochloric acid. Charcoal is too costly and retains too much oil to be used alone. It is usually added at 5–10% by weight of bleaching clay, when required, in order to attain a degree of bleaching not otherwise obtainable.

Not many of the pigments in oil products have been identified although most of them fit into certain categories. The method for their removal depends on the nature of their chemically active groups. Alkalies remove acidic materials. Acid clays remove alkaline pigments.

Xanthophylls, yellow carotenoid pigments containing polar hydroxyl groups, are removed by bleaching clay. Chlorophyll compounds, the green pigment in plant tissues, require acid clay or charcoal for bleaching. Carotene, the major pigment in palm oil and meat fats, is nonpolar and is not removable by adsorption. It is heat bleachable, however, and becomes colorless during deodorization.

Meat fats are sometimes put through a process called bleaching since the mechanics of the treatment are similar. The earth used, however, is a diatomaceous earth or filter aid. It has no adsorptive properties. Meat fats, especially when dry rendered or poorly settled, contain suspended proteinaceous particles which are removed by filtration with the help of the earth. This so-called bleaching is actually a clarification process.

Occasionally bleaching is used to remove colorless matter which could develop color during subsequent processing, especially during deodorization. Residual soaps, gums, nickel compounds from hydrogenation, and partially oxidized fats represent most of this type of material.

As would be suspected, there is also a pigment type that can not be removed by any known means. Cottonseed oil has some unbleachable pigments. These pigments are higher in concentration with expeller oil or oil from seeds which were stored with an excessively high temperature and moisture content. It is possible that gossypol pigment fragments couple chemically with some of the fatty components of the seed under conditions of high heat or high moisture. This is also evident when bleachable cottonseed oil is poorly bleached and then deodorized. Attempts at rebleaching such oil are usually unsuccessful. The color is then said to have been "fixed" or "set" by the heat.

Oils in storage undergo color reversion and sometimes require rebleaching. This is due to oxidation and polymerization of some of the oil components. Golumbic (1943) observed the formation of tocoquinones from oxidation of gamma tocopherol. Bailey (Swern 1964) associated the formation of these colored quinones with color reversion in stored oil.

Deodorizer distillate recovered from soybean oil processing contains pigmented material. The deodorized oil is usually proportionately lighter in color.

HYDROGENATION

Hydrogenation is the treatment of fats and oils with hydrogen gas in the presence of a catalyst. Primarily, the hydrogen reacts with a carbon-carbon double bond and thereby saturates it. Other phenomena take place at the same time with the observable result that the fat or oil becomes physically harder.

Reference is often made to the selectivity of hydrogenation. Oils usually contain a mixture of mono-, di-, and triunsaturated fatty acids. According to one definition, theoretically perfect selectivity in hydrogenation would convert all of the triunsaturated linolenic acid to diunsaturated linoleic acid before it would hydrogenate any of the linoleic to monounsaturated oleic acid. Oleic acid would be hydrogenated to fully saturated stearic acid only after all of the linoleic acid was gone. Complete lack of selectivity would result in hydrogenation of all double bonds on a random basis regardless of on what fatty acids they occurred. Table 12 shows the composition of selectively and nonselectively hydrogenated soybean oil.

In actual practice, hydrogenation of commercial oils approaches a high level of selectivity but falls somewhat short of the theoretical. Nor does physical hardening of the oil result entirely from saturation of a double bond.

If the linolenic acid in a triglyceride such as soybean oil were merely converted to linoleic acid by the saturation of one double bond, the oil would still be entirely liquid at refrigerator temperature. Actually, such oil precipitates some solid fats on being chilled. What has happened is that while one double bond in the polyunsaturated fatty acid chain is being saturated, an adjacent one is being formed into a *trans*-isomer. The *trans*-double bond behaves physically like a saturated single bond. Thus, *cis-trans* linoleic acid behaves somewhat like oleic acid rather than like natural *cis-cis* linoleic acid. The triglyceride containing it solidifies on being reduced in temperature. Similarly, on further hydrogenation linoleic acid is hydrogenated to *trans*-oleic acid which is fairly hard.

Contrary to popular opinion, ordinary margarine oils contain only a trace more of saturated fats than the oils from which they were prepared.

TABLE 12

COMPARISON OF SELECTIVE AND NONSELECTIVE HYDROGENATION OF SOYBEAN OIL

Analysis	Hydrogenation Method	
	Selective	Nonselective[1]
Iodine value	75.8	75.0
FAC melting point °F	90	122
Wiley melting point °F	88	115
Congeal point °F	83	97
SFI at 50°F	33.0	34.4
70°F	18.0	25.6
80°F	10.2	23.1
92°F	1.8	15.1
100°F	0.4	9.1
Fatty acid[2]		
Linoleic	2.2	8.5
Linolenic	0.01	0.57
Oleic	83.7	68.4
Palmitic	10.0	10.0
Stearic	4.1	12.5

Source: Merker (1958).
[1] Phosphoric acid, 0.0067%, added to convertor charge to eliminate selectivity of hydrogenation.
[2] Polyunsaturated acids determined by UV absorption spectrum. Oleic and total saturated acids calculated from this and iodine value. Palmitic acid estimated to yield calculated value for stearic acid.

TABLE 13

FATTY ACID COMPOSITION OF HYDROGENATED SOYBEAN OIL

Fatty Acid	Iodine Value				
	129	107	87	76	5
Palmitic	11	11	11	11	11
Stearic	4	4	5	7	83
cis Oleic	27	27	26	24	0
trans Oleic	0	21	41	52	6
Linoleic	50	34	16	6	0
Linolenic	8	3	1	0	0
Melting point °F	...	81	91	100	145

Source: Weiss (1963).

They are harder because they contain higher levels of *trans*-acids which are, in fact, unsaturated. Table 13 illustrates this phenomenon. As can be seen, the total saturated acids have increased only 2% at an iodine value of 76 while more than $1/2$ of the acids are in the *trans*-form. Ordinary shortenings, however, do contain more saturated fats because they are plasticized by the addition of fully hydrogenated fats.

Low selectivity would result in the formation of fully saturated fatty acids during the course of hydrogenation while maintaining a fair amount of residual linolenic and linoleic acids. Table 12 shows the comparison between SFI values for 2 samples of an oil hydrogenated to the same iodine value under conditions chosen to give high versus low selectivity.

TABLE 14

SFI VALUES FOR SOYBEAN OIL HYDROGENATED TO VARIOUS HARDNESS LEVELS

Iodine Value	SFI Value				
	50°F	70°F	80°F	92°F	100°F
99.3	7	2	0	0	0
90.5	11	5	2	0	0
88.4	15	6	3	0	0
82.5	23	10	4	0	0
78.0	32	16	9	2	0
72.7	40	23	16	5	0
70.0	49	32	25	10	3

Source: From curves in Swern (1964).

The oil hardened under conditions of high selectivity melts sharply. It is similar to a margarine oil in performance. The sample hydrogenated with low selectivity resembles a finished shortening in SFI and melting characteristics. Hard fats formed *in situ* by low selectivity hydrogenation, however, crystallize in the beta phase and are, therefore, undesirable for shortening use. Table 14 gives the change in SFI values with decrease in iodine value for normal hydrogenation.

There are several variations of the hydrogenation process. Most commercial operations are carried out on a batch basis. There are also some continuous hydrogenation installations in use. The basic process in each case calls for intimately contacting gaseous hydrogen, liquid oil, and solid catalyst.

Hydrogenation calls for either bubbling hydrogen through the oil in a heated, agitated, enclosed vessel at atmospheric pressure or adding hydrogen under pressure to the oil in the previously evacuated vessel. Catalyst powder, usually activated nickel metal, is suspended in the oil. Hydrogenation takes place only when the agitator is operating to disperse the gas in the form of a fine froth.

Various processors have developed their own individual hydrogenation techniques. The selectivities obtained depend on conditions used. High reaction temperatures (above 185°C) result in highest selectivity. A lower degree of selectivity, but still a relatively high one, results from using lower reaction temperatures (130°–150°C).

Hydrogen pressures normally used vary from 0 to 50 psi. The pressure has a greater effect on hydrogenation rate than it does on selectivity. Higher pressure means faster hydrogenation.

The nickel catalyst is usually prepared by the reduction of nickel formate suspended in an oil. The oil is saturated with hydrogen as the nickel is reduced so that the catalyst ends up being suspended in a fully hydrogenated fat. The fat is usually flaked for convenience in handling. Most oil processors purchase active nickel catalyst from one of several suppliers. Some still produce their own from raw materials.

Nickel is usually recovered from one batch of hydrogenated oil for reuse in the next batch. Since the catalytic activity of nickel is easily destroyed by one of many types of chemical compounds, reuse of the catalyst requires that oils to be hydrogenated be as free from these catalyst poisons as is possible. In addition, these poisons reduce selectivity of hydrogenation, sometimes to a disastrous degree.

A good general rule is that anything other than neutral oil can act as a catalyst poison. Residual soaps, free mineral acids, sulfur compounds, and carbon monoxide are the major poisons found in actual practice. Soaps can come from poor water wash after refining. Free acid, especially sulfuric acid, can result from leakage of this compound from acid bleaching earth. Carbon monoxide is found in hydrogen gas that has not been properly purified. Careful processing is the best way to minimize catalyst poisoning and thereby obtain the results desired.

Hydrogenation of any oil to any degree of hardness is usually controlled by refractive index (AOCS Method Cc 7-25). This is a simple measurement, rapidly made, with a high degree of accuracy. Refractive index correlates very well with iodine value which is a good general method for expressing hardness of an oil. The relationship of iodine value to hardness depends on the type of oil in question. For example, hydrogenated cottonseed oil has a lower iodine value, perhaps by four units, than hydrogenated soybean oil of an equivalent degree of hardness. The oil processor has already worked out these interrelationships for his own purposes.

When the oil has been hydrogenated to the desired point, the catalyst is removed by filtration, often with the help of bleaching clay and filter aids. Filtration is usually a two-stage process with the "black press" removing the bulk of the nickel and the "white press" giving the oil a final polish.

DEODORIZATION

As the term implies, deodorization is the process whereby the odors and flavors of fats and oils are removed, resulting in a bland finished product. Other compounds are stripped off at the same time. These would include sterols and other waxes, free fatty acids, monoglycerides, some pigments, and fatty oxidation products.

Although deodorized oil is theoretically completely odorless and flavorless, the various types of oils do have characteristic flavors which are distinguishable to the experienced taster. This may be due to traces of residual matter or to oxidation or polymerization of the oil after deodorization even to the slightest degree. The nose is an extremely sensitive instrument and with training can detect compounds that are not measurable by any other means.

By chemical analysis, deodorized oils have a peroxide value of zero as heat destroys these compounds. On storage, oils begin to develop peroxides slowly. If the oils had been badly oxidized before deodorization, they will have poor stability after deodorization. It is suspected that odorless, tasteless oxidation products and fatty polymers have formed prior to deodorization and are not removed by this process. It is also likely that natural antioxidants which are found primarily in vegetable oils have been destroyed by oxidation. They are, therefore, absent from the subsequently deodorized oil. However, addition of fresh antioxidant material to such oil seems to have little effect on improving its stability.

Deodorized oils contain residual free fatty acids (0.02–0.04%) and monoglycerides (0.3–0.5%). It has proved to be impossible to reduce these components to below these levels. The use of steam in the process apparently forms additional free fatty acids and resulting monoglycerides from the triglyceride oils. An equilibrium condition between fatty acid formation and removal rates is attained.

Occasionally water solutions of soaps from the refining process or salts, such as wet sodium phosphate from monoglyceride preparation, pass through filters with the oil to be deodorized. The excess moisture is removed in the deodorization process, causing these extraneous materials to precipitate as a fine cloudy suspension. They should be removed by use of a polishing filter following deodorization.

The process of deodorization calls for blowing steam through hot oil at 200°–260°C under high vacuum (6–12 mm). The steam strips off compounds not otherwise removed. Substitution of nitrogen or hydrogen gas for steam has not proven to be feasible. It is more costly and less efficient than steam.

Deodorizers may be batch, semicontinuous, or continuous in operation. Batch systems are the most widely used although the semicontinuous deodorizer is rapidly gaining in popularity. Continuous systems have not been as widely accepted. They do not allow for easy startup. Most refineries must handle a number of different stocks. Continuous deodorizers are most efficient where only one type of oil is to be deodorized.

The simplest batch apparatus is a large heated vessel with a steam sparger in the bottom and a baffled vacuum takeoff at the top. The oil is usually pumped in relatively cold, heated, stripped, and cooled in the vessel, and then pumped out. The full process requires 10–12 hr for full turnover.

The batch deodorizer has also been designed in three sections. The oil is heated in the first section, deodorized in the second, and cooled in the third. Heat exchangers may be used in multiple systems whereby cold incoming oil is preheated by outgoing hot oil, which obviously is rapidly cooled in the same exchange of heat. Only about 8 hr is required for full turnover.

The semicontinuous deodorizer may be considered as a series of batch deodorizers. The oil is preheated in the top chamber, fed by gravity in sequence to a series of shallow trays containing many bubble-cap steam ports and finally dropped into a collection tank for cooling and discharge. Timing of the process is important. Each movement of oil from tray to tray is carefully controlled by a series of automatically operated valves.

The biggest problem in deodorization is in potential leakage of air into the system. Deodorizers are large with high capacity vacuum pumps. Small air leaks are not easily detected. When they become sufficiently large, the "deodorized" oil darkens and develops a burnt flavor and odor.

The final step in deodorization is the addition of citric acid to the oil. It is usually added as a water solution under vacuum so that the water is evaporated off. The citric acid acts as a metal scavenger, especially for traces of copper and iron which act as prooxidants for oil. The maximum amount of citric acid permitted by U.S. FDA is 0.01%. If citric acid were not added, the oil would revert and oxidize quite rapidly. Fumaric and succinic acids have been suggested as substitutes for citric acid.

Some processors add citric acid solution to oil in the deodorizer prior to the cooling period. Under these conditions, citric acid decomposes to a number of other acids which seem to be equally effective as metal scavengers. It has been suggested that this technique would produce an improved cottonseed oil since citric acid at that temperature would help to decompose the cyclopropenoid fatty acids found in cottonseed oil (Rayner et al. 1966). These fatty acids, occurring at about 0.7% in cottonseed oil, seem to be harmless to man. They have been shown to have a harmful effect on the storage life and hatchability of eggs from hens fed high levels of cottonseed meal.

FRACTIONAL CRYSTALLIZATION

Winterization

Winterization is a specialized form of the overall process of fractional crystallization (Weiss 1967). It was developed to produce salad oil from cottonseed oil. For this reason, winterization has become economically the most important of all fractional crystallization procedures.

When the cottonseed oil market was first being developed, the oil was collected and held in outdoor storage tanks. The oil would solidify in the cold, a characteristic which was found to be annoying and unsightly when it took place in bottled oil on the grocer's shelf. The solid portion of the oil which had set up in storage tanks in the winter at 40°–42°F (it should be remembered that these were southern winters and were therefore relatively mild) was settled out and removed, leaving an oil which would remain clear when chilled. Cottonseed oils were thus divided into *summer* and *winter* oils.

Mayonnaise was increasing in popularity at about the same time. It was not a commercial product at first but was prepared for immediate use by the chef or housewife. The demand for a commercially produced mayonnaise led to increased need for oils which would not solidify on chilling and thereby break the mayonnaise emulsion. Winterized cottonseed oil was the natural answer. It thus became known as a *salad* oil to distinguish it from the unwintered oil which was then called *cooking* oil.

The demand for salad oil could no longer be filled by waiting for winter. The simple expedient was undertaken of storing the oil in quiescent indoor tanks and applying winter conditions by chilling the entire room. This process is still used today by a number of winter oil producers. Other processors use jacketed tanks with or without agitation.

The stearine crystals usually separate in soft, flat plates. These are difficult to filter as they tend to blank off or seal the filter cloths if excess pressure is applied to the oil stream. In many cases the oil is fed to a plate and frame press by gravity.

Where agitation is used in a winter cell, it is slow and gentle. It may even be stopped once crystallization has begun. Undue agitation of the crystalline mass tends to break it up into a slimy mass which is virtually unfilterable. Centrifugal separation has never been particularly successful. The mechanical action of pumping the crystals to the centrifuge and the action of the centrifuge apparently packs the crystals in such manner that they entrain an excessive amount of oil.

Certain chemical compounds have been developed as crystal inhibitors to retard formation of solid fats in salad oils, in short, to extend the cold test. These compounds will be discussed more thoroughly in Chap. 4. They also function as crystal modifiers in winterization (Anon. 1968). Their action is to change the shape of the stearine crystal so that the crystalline mass would retain less oil and permit the oil to filter at a faster rate.

Winterization from whole oil is a slow process. Cold oil is viscous so that the growth of the stearine crystal to maximum size would take a long time. Roughly 2–3 days are required from start of the chilling operation to the end of filtration of a single batch.

Solvent winterization is a relatively recent development (Cavanagh 1959; Cavanagh *et al.* 1961). In this process, the oil viscosity is reduced by dissolving it in a solvent such as hexane. The stearine crystals form in a few minutes and can be separated readily with a minimum of entrainment of the low-viscosity liquid phase. The stearine mass is relatively hard.

Lightly hydrogenated soybean oil (105 iodine value) is winterized into salad oil in the same manner as is cottonseed oil.

Stearine Pressing

Certain semihard fats, such as lard and tallow, were fractionated by solidifying the fats at a specific temperature and pressing the material to separate it into a stearine and an oil portion. The chilling operation was carried out by filling small tanks, usually on wheels, with melted fat. The tank carts were lined up in a temperature-controlled room and held for 3 to 4 days. The normal temperature for lard was about 50°F, for oleo stock, about 84°–90°F. The solidified fats were then wheeled to the press room where they were shoveled by hand into large canvas cloths. The filled press cloths were folded and placed in a frame press mounted vertically over a tank. Pressure was applied so that the oil was squeezed into the tank. After a sufficient length of time for drainage, the pressure was released and the stearine cakes were dumped by hand from the unfolded press cloths into a melting tank.

Palm kernel oil is being processed in a similar manner at this time for use in confectioners' pastel coatings.

Production of fat fractions for use in confectioners' cocoa coatings is carried out by crystallization from solvent. Palm oil, tallow, partially hydrogenated cottonseed or soybean oils among other oils are dissolved in hexane and chilled to remove the trisaturated triglycerides. These are probably incorporated into shortenings as hard stock. The solution is then chilled further to precipitate the disaturated triglycerides. These are freed from solvent to become the desired hard butter or confectioners' coating fat. The residual fats, when freed from solvent, are useful as high-stability frying oils which are liquid at room temperature.

Similar frying oils are produced by fractional crystallization without solvent (Simmons et al. 1968). The solids are removed by filtration or centrifugation but without pressing. In this case, the oil is the desired product and the hardness of the solids is of no real consequence. The solids contain a sufficient amount of entrained liquid oil to make them useless as a confectioners' hard butter.

INTERESTERIFICATION

An illustration of interesterification was given in Chap. 1 showing the fatty acid distribution resulting from randomization of a mixture of tripalmitin and triolein. This idealized situation is not too far removed from practical considerations of the interesterification reaction.

Fully hydrogenated cottonseed oil is interesterfied with fully hydrogenated coconut oil for the commercial production of hard butters. No attempt will be made to calculate the various triglyceride types resulting from this reaction, but there must be a very large number of them. Randomization of just 3 fatty acids gives 10 basic triglycerides with a total of 17 isomers.

Interesterification does not require that two separate fats be mixed to be randomized. A single fat such as lard is usually interesterified by itself. The untreated fat must contain specific triglycerides with fatty acids in a nonrandom distribution. In the case of lard, palmitic acid occurs in the 2-position in the glyceride molecule. The uniformity in position of a fatty acid favors the formation of large beta crystals. A mixture of glyceride types inhibits large crystal growth, favoring development of small beta-prime crystals. Interesterification brings about random distribution and thereby alters the physical characteristics of the fat. In the case of lard, the SFI values and crystal structure both change markedly. Figure 10 (Chap. 2) shows the SFI curves involved. As was pointed out under lard in Chap. 2, the beta crystal structure of lard changes to beta-prime with interesterification.

Directed, or nonrandom interesterification was developed in which the reaction is carried out at a sufficiently low temperature to cause precipitation of trisaturated triglycerides during interesterification. The major purpose of this method, as far as lard is concerned, is to provide sufficient hard fats from the lard itself to aid in plasticizing it into a finished shortening (Hawley and Holman 1956). Lard is processed by directed interesterification on a commercial scale for shortening manufacture.

Interesterification is carried out by adding an appropriate catalyst to the fat at the proper temperature. The fat is agitated until the reaction is judged to be complete. The catalyst is then inactivated and removed.

The first catalysts used were tin salts or sodium soaps. The temperatures required were very high, 225°–250°C. It was then learned that sodium methoxide, sodium metal, or sodium-potassium metal alloy would operate at low temperatures, from 5° to 135°C, and at much higher rates of activity. Random interesterification is usually carried out at 95°–135°C and the directed reaction at 32°–38°C (Hawley and Holman 1956).

The end point of lard interesterification is best judged by some form of cooling curve (Fig. 6 and 7) which indicates the absence of the heat of crystallization associated with untreated lard. The end point of interesterification for hard butter production might be determined by loss of the rapid solidification of the fully saturated glyceride component on cooling a small sample.

The catalyst can be destroyed in open kettle reactions by the addition of a small amount of water. With this method, the catalyst forms a soap which can be settled out and separated as in kettle refining. If interesterification is conducted in an enclosed tank, it is better to destroy catalyst activity with phosphoric acid. The resulting sodium phosphate is easily removed by filtration.

The process is usually finished by bleaching to remove traces of soap, salts, and color. Fats usually develop a brownish pigmentation during the reaction due to the activity of the catalyst. Deodorization is then required to remove free fatty acids and, where sodium methoxide is the catalyst, an accumulation of methyl esters of fatty acids.

GLYCEROLYSIS

Glycerolysis refers to the reaction of triglycerides with an additional amount of free glycerin. The resulting product is a mixture of mono- and diglycerides in which the glycerin molecule is esterified with only 1 or 2 fatty acids. The amount of monoglyceride obtained depends on the proportions of glycerin and triglyceride used. The reaction mixture most commonly used yields a product containing 40–42% monoglycerides. They can be analyzed chemically by titration with periodic acid reagent (AOCS Method Cd 11-57).

Glycerin is not very soluble in fats until the temperature of the mixture reaches about 200°C. It is therefore necessary to carry out the preparation of monoglycerides at 200°–250°C. The reaction then goes very quickly. The catalyst used is sodium hydroxide. Phosphoric acid is used to kill the catalyst. The reaction mixture is then cooled. Excess glycerin containing the phosphate salt residue is drained off. The monoglyceride mixture is given a light vacuum stripping to remove any unsettled glycerin. Deodorized fats are usually used for the reaction. Monoglycerides are sufficiently volatile to be removed during subsequent deodorization.

A specialized monoglyceride product is commercially available. It is called a self-emulsifying monoglyceride. In this product, the catalyst is not inactivated with phosphoric acid. The reaction mixture is cooled quickly after reaching equilibrium. The catalyst becomes inactive at the reduced temperature. Some of the glycerin begins to precipitate before the catalyst reaches inactivation temperature, causing the equilibrium to shift towards lower levels of monoglyceride. The monoglyceride content is about 32% with about 7% free glycerin. On the other hand, the soap content is in the order of 2%, since the catalyst has not been neutralized. The mixture makes a highly active emulsification system.

MOLECULAR DISTILLATION

High molecular weight products can be distilled successfully to attain a high degree of purity. The vacuum is measured in microns. The material to be distilled is spread on a heated surface in a thin layer and the receiving cold surface is located close to the heated surface. This provides a relatively short path for travel of the volatilized material.

Distilled monoglycerides of 90% purity by analysis, tocopherols, and vitamin A are the major fatty products prepared by molecular distillation.

PLASTICIZING

While a number of special products will be discussed later in terms of specific processing and packaging requirements, shortenings are generally plasticized and packaged by a standardized system (Joyner 1953).

Plasticizing is carried out by means of a scraped-surface internal chilling machine. The most commonly used machine of this type is known as

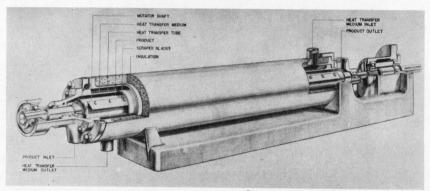

Courtesy of Votator Div., Chemetron Corp.

Fig. 11.   Votator A-unit

the Votator. It is divided into "A" (chilling) units and "B" (crystallizing) units. The A-unit (Fig. 11) consists of a cylinder containing a coolant, usually compressed ammonia, surrounding an internal cylinder containing the material to be chilled. The chilled cylinder wall is scraped continuously by a revolving blade attached to a center shaft. The melted shortening or margarine emulsion is pumped through the inner cylinder where it is supercooled and begins to crystallize. The still fluid fat is then piped to a B-unit for solidification.

There are two types of B-unit, a dynamic unit for shortenings (Fig. 12), and a quiescent unit for table-grade printed margarines. The dynamic unit consists of a shaft with a double row of pins protruding from it mounted inside a cylinder. A row of pins on the cylinder wall alternates with the pins on the shaft. The shaft is rotated at high speed as the supercooled fat passes through it. The fat begins to solidify in the B-unit and is thoroughly whipped by the action of the revolving pins. The crystallized but fluid product is then filled into a container and set aside. The crystallized mass solidifies shortly after it is permitted to rest without movement.

Bulk margarines are chilled in the same manner. The only difference between shortenings and bulk margarines as far as chilling is concerned

Courtesy of Votator Div., Chemetron Corp.

FIG. 12. VOTATOR DYNAMIC B-UNIT

is that air or nitrogen is usually incorporated into melted shortenings just prior to their entering the A-unit. Margarines contain water or milk in emulsion form and are not normally aerated. Aeration gives shortenings their whiteness and opacity. The aqueous phase of margarine emulsions has the same effect as air.

The quiescent B-unit is used to give the chilled but fluid table margarine emulsion time to solidify but without being whipped. The B-unit is usually split lengthwise into two sections and has a perforated end plate to provide back-pressure. A reciprocating valve opens one section first, closing the second. The fluid oil pushes the solid emulsion out of the first section. The valve is timed to switch at this point to close the first section and open the second. The chilled fluid margarine in the first section now has the opportunity to solidify by resting. The second section is emptied of solid margarine and replaced with fluid emulsion. This action alternates repeatedly between the two sections. The solidified margarine is pushed from the B-unit through the perforated plate as noodles which are then compressed into prints. This will be described further in Chap. 9 on margarine.

The chilling of shortening and margarine products is part art and part science. If the chilling machine operator attempts to fill his product by temperature and pressure alone, he will be unable to obtain a uniform texture in the product. The shortening or margarine should be filled in such a manner that it forms a slightly rounded surface. Too warm a fill will be soupy and show a flat surface. The finished shortening will be too hard and unworkable at low temperatures. Too cold a fill will result in excessive mounding in the container and a hard and lumpy finished product. The temperature at which the correct appearance is obtained should fall within a definite predetermined range for the given product. If it does not, the indication is that the formulation was in error.

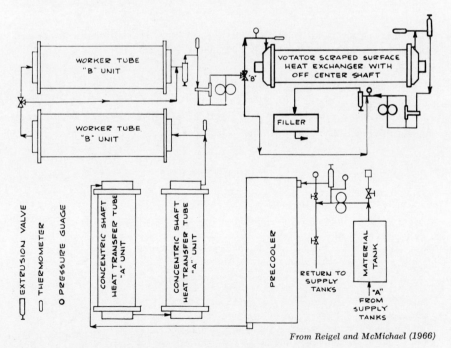

From Reigel and McMichael (1966)

FIG. 13.   SHORTENING PLASTICIZING SYSTEM WITH POST-CHILLING

Reprinted from *Journal of the American Oil Chemists' Society*

TEMPERING

Tempering of finished shortenings is a phenomenon which can often only be demonstrated by performance testing. In some cases, consistency or penetration values undergo some change during tempering, usually showing a softening of the shortening product.

Specifically, a freshly filled shortening will not cream properly in the baking of cakes or production of icings. Its creaming properties will improve to a maximum value after 2–4 days of tempering. All that is required is the holding of the packaged shortening at a temperature close to that at which the shortening was originally filled (80°–85°F). Tempering under these conditions merely happens. Shortening plants in southern U.S. locations often obtain tempering unwittingly.

If a recently filled shortening is placed in an excessively cold location it may become permanently hard and brittle. Warming it to "tempering" conditions will have no effect on it even after long storage time.

Attempts have been made to shorten tempering time. Shortenings may be tempered quickly if filled by chilling in a normal A- and B-unit sequence followed by a second A-unit (Reigel and McMichael 1966). A schematic drawing of the system is given in Fig. 13. It is postulated that the second A-unit removes excess heat of crystallization. The physical work under these conditions hastens tempering. It also results in a more uniform fill from batch to batch (Feuge et al. 1962).

Tempering seems to be related to some crystal phenomenon. Attempts at observing this through X-ray diffraction or microscopic studies have shown very little change in crystal structure when a well chilled shortening becomes tempered. A good scientific explanation of tempering of a shortening has not yet been found.

BIBLIOGRAPHY

ANON. 1968. Cuts filtration time after winterizing. Food Process. 29, No. 12, 47.

BATES, R. W. 1968. Edible rendering. J. Am. Oil Chemists' Soc. 45, 421A–422A, 424A, 430A, 462A, 464A.

CAVANAGH, G. C. 1959. Method of separating components of cottonseed oil by fractional crystallization of miscella thereof. U.S. Pat. 2,883,405. Apr. 21.

CAVANAGH, G. C., CECIL, E. J., and ROBE, K. 1961. Seed to salad oil in 18 hours. Food Process. 22, No. 4, 38–45.

DOWNING, F. P. 1959. The production of meat and fat products through centrifugal rendering. J. Am. Oil Chemists' Soc. 36, 319–321.

FEUGE, R. O., LANDMANN, W., MITCHAM, D., and LOVEGREN, N. V. 1962. Tempering triglycerides by mechanical working. J. Am. Oil Chemists' Soc. 39, 310–313.

GOLUMBIC, C. 1943. The autoxidative behavior of vegetable and animal fats. Oil Soap 20, 105–107.

HAWLEY, H. K., and HOLMAN, G. W. 1956. Directed interesterification as a new processing tool for lard. J. Am. Oil Chemists' Soc. 33, 29–35.

JOYNER, N. T. 1953. The plasticizing of edible fats. J. Am. Oil Chemists' Soc. 30, 526–535.

MERKER, D. R. 1958. Non-selective hydrogenation of fats and oils. U.S. Pat. 2,862,941. Dec. 2.

RAYNER, E. T., BROWN, L. E., and DUPUY, H. P. 1966. A simplified process for the elimination of the Halphen test response in cottonseed oils. J. Am. Oil Chemists' Soc. 43, 113–115.

REIGEL, G. W., and MCMICHAEL, C. E. 1966. The production of quick tempered shortenings. J. Am. Oil Chemists' Soc. 43, 687–689.

SIMMONS, R. O., REID, E. J., BLANKENSHIP, A. E., and MORGAN, P. W., JR. 1968. Production of liquid shortening. U.S. Pat. 3,394,014. July 23.

SWERN, D. (Editor). 1964. Bailey's Industrial Oil and Fat Products, 3rd Edition. Interscience Publishers Div., John Wiley & Sons, New York.

WEISS, T. J. 1963. Fats and oils. In Food Processing Operations, Vol. 2, M. A. Joslyn and J. L. Heid (Editors). Avi Publishing Co., Westport, Conn.

WEISS, T. J. 1967. Salad oil manufacture and control. J. Am. Oil Chemists' Soc. 44, 146A, 148A, 186A, 197A.

CHAPTER 4

Chemical Adjuncts

A large number of chemical compounds exist which have specific function in shortenings of various types. Most of these additives can be classified into several categories, namely, emulsifiers, stabilizers or antioxidants, metal scavengers, antifoam agents, crystal inhibitors, preservatives, pigments, and flavors.

EMULSIFIERS

Emulsifiers represent the largest group of individual compounds. This is because they each exhibit a particular functionality and except for occasional instances, are not completely interchangeable with each other.

Attempts have been made to grade emulsifiers on a systematic scale so as to be able to predict activity and thereby potential utility. The HLB (Hydrophilic-Lipophilic Balance) system has been only partially successful in this. The HLB number is an expression of the ratio between the hydrophilic, or water soluble, and the lipophilic, or fat soluble, portions of the emulsifier system. The system may consist of a single compound or, preferably, a mixture of two or more emulsifiers. The scale runs from 1 to 20 with 10 indicating an equal amount of hydrophilic and lipophilic groups (Knightly and Klis 1965). Table 15 lists HLB values for a number of emulsifiers commonly used in shortenings.

The ability of an emulsifier to disperse in water can serve as a rough guide to HLB value. Products with a value below four do not disperse in water. Those with a value above ten disperse well and may be water soluble. Emulsifiers that lie between these limits form milky dispersions with various degrees of dispersive stability (Griffin 1965).

HLB values can help in screening an emulsifier. Cake baking, for example, would require an emulsifier with a value between 2.8 and 4.0. No chemical method is available for rating specific emulsifiers falling within this range. Emulsifiers with the same HLB value may vary from excellent to poor in their ability to produce an acceptable cake. The effect of the particular emulsifier on starch hydration is important. HLB value does not measure this attribute.

Performance testing is the only solution to the problem of selecting an appropriate emulsifier or emulsifier blend. This is not a simple task as concentration of the emulsifier in the product to be evaluated is also a factor. In cake baking, it is often difficult to tell whether a cake is under-

TABLE 15

HLB VALUES OF VARIOUS EMULSIFIERS

Emulsifier	HLB Value
Mono- and diglycerides	2.8–3.5
Glycerol monostearate	3.8
Lactylated mono- and diglycerides	2.6
Propylene glycol monostearate	3.4
Sorbitan monostearate	4.7
Polysorbate 60	14.9

Source: Griffin (1965).

or overemulsified. It is necessary to run such tests in a series in which emulsifier is added to the shortening in increasing levels of concentration.

A complicating factor arises from the observation that most, if not all, cake emulsifiers work best when used in conjunction with monoglycerides. A few that are apparently not enhanced by added monoglyceride are found to have an almost optimum amount of this component developed in them during their chemical synthesis. Development of properly emulsified shortenings, therefore, requires a complex performance testing program in which both the new emulsifier and specific monoglycerides are varied simultaneously. Statistical patterns have been postulated for such programs (Truax and MacDonald 1960; MacDonald *et al.* 1966).

Care must be taken in selecting the proper emulsifier increments. A cake formulation may be sensitive to 0.25% increments of analyzable monoglyceride in a shortening.

A testing program which uses 1% increments would completely miss finding the optimum emulsifier level. Emulsifier selection is obviously a long, involved, and tedious task.

Development of new synthetic emulsifiers is not as haphazard as may seem from the above. The basic concept is to couple a hydrophilic compound to a lipophilic compound (MacDonald 1968). The hydrophilic group is usually selected from compounds containing a large number of oxygen atoms, either as hydroxyl, carboxyl, or similar groups, or as ether linkages. Many compounds contain a mixture of these. Nitrogen and sulfur compounds might be more active. They are rarely, if ever, used for edible synthetic emulsifiers as they often contain toxic residues. Lecithins and lecithoproteins are nitrogen-bearing emulsifiers but they are obtained from natural sources. The lipophilic group is normally a fatty acid.

Lecithin

Crude lecithin is used as an emulsifier in many food products. Various purified and processed grades are also available. Ordinary lecithins are marketed as unbleached (UB), single bleached (SB), and double bleached (DB). Bleaching is usually accomplished by treatment with hydrogen peroxide. These products do not differ from each other in emulsifier function. Unbleached lecithin is satisfactory for use in dark products such as chocolate. The bleached products are required for uses where dark colors may be objectionable.

Lecithin is available as a regular waxy solid or as "fluid," a molasses-like syrup. Fluidization is attained by addition of 2–5% fatty acids. More recent developments have achieved fluidity by mechanical means. The choice of which product texture to use in this case is based on mechanical handling requirements by the user. Regular and fluid lecithins have equal emulsification properties.

Lecithin derivatives are available for specialized uses. These include alcohol extracted lecithins and hydroxylated lecithins. They are reputed to have enhanced emulsification properties. This is for the potential user to determine by performance tests with his own particular product.

Lecithoproteins are lecithin derivatives. They are highly active naturally occurring emulsifiers. They are found in egg yolk, skim milk and defatted cream solids, wheat flour, and similar sources. Unfortunately, they have never been isolated or synthesized on a commercial scale.

Monoglycerides and Derivatives

The preparation and chemical structure of mono- and diglyceride products have been discussed in Chap. 3 under "Glycerolysis." They are prepared from many types of fats and oils since the hardness of the oil, as indicated by iodine value, has a real effect on the functionality of the monoglyceride prepared from it.

The value of the diglyceride component as an emulsifier has been the subject of much debate but practically no scientific evidence. A series of photographs of cakes baked with varying proportions of mono- and diglycerides has been published by Swanson (1955). This work implied that diglycerides were of value in cake baking. Unfortunately, no methodology nor numerical data was included to aid in interpreting the findings.

Distilled monoglycerides containing 90% active material are also available in an assortment of fatty acid sources from unhydrogenated vegetable oils to fully hydrogenated fats. Here too, functionality and utility depend on the hardness of the fatty acids.

It is an interesting concept that monoglycerides are considered as chemicals by legal definition. In many instances, "all vegetable" products may contain monoglycerides from meat fat sources.

The emulsification properties of monoglycerides have been altered by coupling them with more hydrophilic compounds. These derivatives have found a number of uses. Reaction of monoglycerides with lactic acid has been thoroughly exploited. This type of product was first described by Tucker (1943). He referred to higher fatty acid esters of a water soluble polyhydric alcohol esterified with a water soluble hydroxycarboxylic acid. A number of compounds fit this broad description. The major commercialization, however, was based on reaction products of palmitic acid, glycerin, and lactic acid (Little 1949; Chang et al. 1960; Radlove et al. 1960; Shapiro 1961; Woods 1961). The many process variations described probably all yield the same equilibrium product.

Sulfoacetate and diacetyl tartaric acid derivatives of monoglycerides have found some utility as emulsifiers, the former in margarine, the latter in bread (Ziemba 1966). The reaction between monoglycerides and citric acid has been developed as a means of solubilizing citric acid for use as a metal sequestering agent (Hall 1957). By adding propylene glycol to the monoglyceride-citric acid reaction mixture to act as a mutual solvent, a new active emulsifier was formed which proved to be mixed esters of glycerin, propylene glycol, citric acid, and fatty acids (Kidger 1962).

Propylene Glycol Esters

Propylene glycol monostearate (PGMS) was found to be functional as an emulsifier for cakes. The first cakes had to be low in fat. The shortening required an unusually high level of PGMS, however. The amount incorporated was 15 to 20%. It was then found that mixtures of about equal amounts of PGMS and monoglycerides were more functional than PGMS alone. The mixtures were similar in activity to the lactylated monoglycerides (Kuhrt and Broxholm 1962A, B, 1963; Kuhrt et al. 1963).

Polyglycerol Esters

Glycerol can be polymerized by heating it under vacuum with a catalyst such as sodium acetate. The polymer chains are 2 to 12 units in length with average values of 3, 6, or 10 units of glycerin. The polyglycerol molecules are then esterified with various fatty acids in different proportions to give a variety of average ratios of fatty ester groups to free hydroxyl radicals.

Polyglycerol esters have been studied in a number of use situations (Nash and Knight 1967; Babayan 1968A, B). They have been found to help retain gloss in chocolate, prevent oil separation in peanut butter, extend cold test of salad oil, reduce spattering in margarine, and emulsify cake and icing shortenings. Their versatility is evident from the listings of HLB values for the various polyglycerol esters (Table 16). Polyglycerol compounds have a characteristic caramelized odor and flavor.

Sorbitan and Polysorbate Esters

Reaction between sorbitol and fatty acids results in a series of sorbitan esters. These are then condensed with ethylene oxide to give a parallel series of polyoxyethylene sorbitan esters (Griffin 1965). The term "polysorbate" has been coined for the sorbitan esters condensed with 20 units of ethylene oxide. The monostearate esters, sorbitan monostearate and

TABLE 16

HLB VALUES FOR POLYGLYCEROL ESTERS

Decaglycerol Ester	HLB Value
Monolaurate	12–14
Distearate	7–9
Tristearate	6–8
Tetraoleate	5–7
Octaoleate	3–5
Decaoleate	2–4

Source: Nash and Knight (1967).

polysorbate 60, are the most important of these for fat and oil products.

Polysorbates are extremely active as emulsifiers. They are useful at much lower levels than required for other emulsifiers. Polysorbates are also quite bitter, which can be a problem in bland foods.

Sucrose Esters

Sucrose has eight hydroxyl radicals which can be esterified with fatty acids. The potential range of products, from mono- through octaester, is large. The HLB value for sucrose monopalmitate is given as 14, for the dipalmitate, 7 (*Nitto Ester Technical Information Bulletin*, Dai-Nippon Sugar Mfg. Co., Tokyo, Japan). Sucrose monopalmitate is similar to polysorbate 60 in activity.

Sucrose esters are currently prepared by reacting sucrose and methyl palmitate or stearate with sodium methoxide or potassium carbonate catalyst. The reaction is carried out in dimethylformamide as the solvent (Osipow *et al.* 1956; Hass *et al.* 1959). The products of this process have not been approved for use as a food additive in the United States as they contain traces of dimethylformamide or other nitrogenous residues, even after extensive purification.

Work is now in progress to develop a reaction system which will produce sucrose esters without residual toxic compounds (Babayan and Atikian 1960; Osipow and Rosenblatt 1967; Feuge *et al.* 1970).

Lactic Acid Emulsifiers

Lactic acid is used in the preparation of specialized emulsifiers in addition to the lactylated monoglycerides. One of these is the reaction product between tallow fatty acids and beta-propiolactone. It is known as tallowyl beta-lactic acid (Young and Spitzmueller 1960). Similar products are made by reacting stearic and lactic acids (Wilke 1968) or stearic and fumaric acids (Anon. 1966). Sodium and calcium stearyl-2-lactates and sodium stearyl fumarate are used as bread tenderizers.

ANTIOXIDANTS

The term "stabilizers" when applied to fats and oils refers to antioxidants. Tocopherols are natural antioxidants found in vegetable oils. The addition of synthetic antioxidants has little effect on extending the shelf-life of vegetable oil products and is rarely used in them (Pohle *et al.* 1964). Safflower oil is one exception to this. Unhydrogenated safflower oil will become rancid rapidly if not protected by antioxidants.

Antioxidants were developed primarily for use with meat fats. Originally, small amounts of cottonseed oil (as a source of tocopherols), lecithin, and gum guaiac were used as stabilizers for lard. Modern antioxidants for food fats are propyl gallate (PG), butylated hydroxyanisole (BHA), and butylated hydroxytoluene (BHT).

There are a number of antioxidant preparations on the market containing the above compounds in various proportions and in various solvents such as vegetable oils or propylene glycol. The antioxidant compounds have a synergistic effect on each other. A combination of two or more of them at any given total concentration is more effective in extending shelf-life of the fat than any one of the antioxidants alone at the same level of use.

The U.S. FDA under the Food Additives Amendment permits the use of the various antioxidants at a maximum level of 0.01% of any one and a maximum of 0.02% of any combination of antioxidants in a shortening product.

The various antioxidant compounds have different individual characteristics. PG develops a blue color by reacting with iron in the presence of moisture. Metal sequestrants such as citric acid seem to prevent this. Complete absence of moisture will also prevent blue color formation. PG is volatile and is lost in baking and frying. Thus it will protect a shortening in storage but will not protect foods baked with or fried in the shortening. PG gives shortenings in which it is used unusually high AOM values. There is evidence, however, that the high AOM in this case does not reflect an equally high shelf-life for that shortening (Pohle *et al.* 1964).

BHA and BHT do not develop off colors. They are at least partially retained in frying and baking. They do stabilize foods fried in or baked with the shortenings containing them. BHA and BHT are especially effective in combination with each other. They have been shown to extend the actual shelf-life of shortenings containing them beyond the time expected from AOM values. On frying, shortenings containing BHA tend to give off a strong phenolic odor. For this reason, it has been expedient to reduce the amount of BHA in frying fats to about $1/2$ that permitted by the FDA. BHT does not seem to give off sufficient odors on frying to be objectionable.

The apparent action of the antioxidant is to use up available oxygen in the shortening without developing rancid odors and flavors by oxidizing the fat being protected. Antioxidants are analyzed by measuring their reducing action. Oxidized antioxidants have no reducing properties. When an antioxidant is first added to a shortening, analysis can detect practically all of the amount added. On storage of this shortening for some length of time, analysis can often detect only $1/2$ or less of the added antioxidant, showing that the antioxidant has been oxidized.

METAL SCAVENGERS

Metal scavengers or sequestering agents have the function of removing and thereby inactivating trace metals in the fats or fatty products. The most damaging metals are copper and iron which are strong prooxidants for fats. Traces of copper are sufficiently active to completely off-set the activity of added antioxidants.

The most commonly used metal sequestrant is citric acid. Isopropyl citrate, stearyl citrate, and monoglyceride citrate are also used. These compounds are usually added directly to the fats.

Ethylenediamine tetraacetate (EDTA) salts, e.g., the calcium disodium salt, are powerful sequestrants which are normally added to fatty foods containing an aqueous phase. Margarine, mayonnaise, and salad dressing emulsions are good examples of such foods. The EDTA salts are supposedly required to remove trace metals entering these foods through salt, spices, and similar ingredients (Melnick 1961; Melnick and Ackerboom 1961; Stapf 1959).

ANTIFOAM AGENTS

When food is fried in fats, moisture in the food expands and is released in masses of bubbles. Some components in the food material or in the fats, especially in old and somewhat degraded fats, may cause retention of the moisture in the form of a foam. Normal bubbles of moisture are large and break readily. Foam bubbles are small and seem to linger. They may flood the fry kettle if the foam layer becomes thick enough.

Antifoam compounds aid by retarding foam formation. They are mixtures of organic silicon oxide complexes dispersed on colloidal silica. They are called dimethyl siloxanes or, more commonly, silicones. Silicones are added to frying fats at levels of 0.5–3 ppm (Martin 1953). Higher concentrations do not inhibit foam formation more effectively nor do they extend the frying life of the shortening to a greater extent. Surprisingly, 50–100 ppm silicone levels may even cause foaming of a frying fat which would not have foamed ordinarily.

The effectiveness of silicone antifoam agents can easily be demonstrated by frying small controlled amounts of food in identical shortenings, one with, the other without added silicones. The shortening without silicones will develop foam after frying only a few batches of food. The treated shortening will fry 5 to 10 times as many batches before showing tendency to foam.

Since cake batters are also foams, the unintentional presence of silicones in cake shortening can cause failure of the cake to reach the desired volume.

CRYSTAL INHIBITORS

Cottonseed and hydrogenated soybean salad oils tend to deposit fatty crystals in the refrigerator. The length of time required for the first appearance of these crystals can be increased by the use of fatty crystal inhibitors (Weiss 1967). From the viewpoint of the oil processor, crystal inhibitors lengthen the cold test of the oil.

Chemically, the crystal inhibitor is a fat soluble product similar to a triglyceride in general structure but differing from it in some specific way. Since crystals grow by the orderly deposition of similar molecules, the inclusion of a dissimilar molecule can stop crystal growth as long as the different molecule remains in place in the crystal lattice. The dissimilar molecule, or crystal inhibitor, must have some similarity to the other molecules present or it would not have deposited on the crystal surface in the first place.

Lecithin was the first crystal inhibitor used (Grettie 1936). It is a triglyceride with one fatty acid replaced by a phosphoric acid ester of choline or ethanolamine. There are a number of chemical compounds which have been developed for use in controlling crystal growth but only two have been approved for use by the U.S. FDA under the Food Additives Amendment. The first of these is called oxystearin. It is prepared by heating fully hydrogenated cottonseed or soybean oil to a high temperature while blowing air through the hot oil. The exact chemical composition is not known but it contains many polymers and breakdown products of triglyceride molecules.

The other approved crystal inhibitor is a polyglycerol ester of various mixed fatty acids. In this case the fatty acids are normal but the glycerol portion of the triglyceride is replaced by glycerol polymers.

The efficacy of the crystal inhibitor depends on the oil to which it is added. A poorly winterized oil cannot be made into a good salad oil by the use of these additives. A low cold test can be extended to only a small degree, e.g., a 5-hr test can be lengthened to perhaps 10 hr (Nash and Knight 1967; Anon. 1968). A good starting oil, however, can be made into a superior oil with the use of a crystal inhibitor. A 15-hr untreated oil can become an 80-hr finished salad oil.

PRESERVATIVES

The term preservative will be used in this discussion as a material which will preserve a product from microbial spoilage. Unfortunately, the term has been used for antioxidants, antifoam agents, and other compounds which retard change in a product. Spoilage from bacteria, mold, and yeast contamination is a problem in fatty foods which contain moisture. Mayonnaise and margarine are the major products of this type.

Salt and vinegar are excellent preservatives although they are usually added to foods as seasonings. The salt level in regular margarine is 2.5–3.5%. Since the moisture level is only about 15%, the salt concentration in the aqueous phase is 13–19%. This is sufficient to protect margarine from spoilage.

Sodium benzoate, benzoic acid and potassium sorbate are permitted as preservatives in margarine under U.S. FDA Standards of Identity. Soft margarines which would be excessively salty to taste at salt levels of more than 2% require additional protection. Benzoates and sorbates are sometimes used together as sorbic acid is most effective in retarding growth of yeasts and molds while benzoic acid works best against bacteria. Both of these preservatives are most effective at a low pH range of 3 to 4. However, they are sufficiently active at the pH range of 5 to 6 found in modern margarines.

Vinegar is the sole preservative permitted in mayonnaise, French, and spoonable salad dressings under U.S. FDA Standards of Identity. The concentration of acetic acid in the aqueous phase is important. It should be at least 2.5% for adequate protection. The salt content of such products is too low to be of any consequence as a preservative.

PIGMENTS

Pigments used for fat and oil products are usually oil soluble and in the yellow to reddish-orange range. While it is possible to use lakes for opaque products such as margarine, this is not usually done.

The carotenoids are practically the only oil soluble colors approved for use by U.S. FDA. These include the carotenes, bixin, and apo-6-carotenal.

The carotene pigments were once obtained entirely from natural sources. Most of these have been replaced in recent years by synthetic beta-carotene. The synthetic product is used extensively in table margarines.

Oleoresin paprika is an excellent source of natural pigments. It is used in mayonnaise and salad dressings. It is both excessively red in color and too high in characteristic flavor to be used in margarine.

Bixin and curcumin are not ordinarily available as pure compounds. They are the primary pigments found in two natural materials. Bixin, a carotenoid, is the main coloring matter in annatto seed. The seed is ground and extracted to produce a reddish-orange pigment. Curcumin, a phenolic diketone, is the greenish-yellow pigment found in turmeric root. When annatto and turmeric extracts are blended, the resulting pigment can vary from a pure yellow to an orange-yellow depending on the proportions of the two materials. A blend resembling beta-carotene in the hue imparted to margarine is used for this product, both for table and bakery grades (Todd 1964).

Carotenoids are destroyed by heat such as is used for frying. Oil suspensions of carotene and solutions of annatto are stabilized with BHA and BHT to be sold as heat-stable pigments for frying oils and popcorn oils. These pigments will bleach in deep fat frying even when stabilized. They are sufficiently stable for coloring pan-fried foods and popcorn for which the heat-stable pigment solutions were designed.

Apo-carotenal is a synthetic pigment which is used in oil primarily as a color intensifier for beta-carotene. It enhances the red tones of carotene, making it undesirable for some uses which require more of a yellow hue.

FLAVORS

The flavors used in margarine, some shortenings, and cooking oils are described as butter-like. At one time, diacetyl was thought to be the only flavor in butter. It was, therefore, the only flavoring compound permitted in margarine. Diacetyl was considered to be naturally occurring if it was developed during the culturing of specific bacteria in milk. The bacterial inoculum was called a starter.

Cultured milk was distilled to obtain diacetyl. The resulting product was called a starter distillate. When this or a chemically synthesized diacetyl was added to margarine, the margarine was considered to be artificially flavored.

Development of improved analytical techniques led to the finding of other flavor components in butter. These include butyric and other short-chained fatty acids and various long-chained lactones. The Standards now permit the addition to margarine of any safe compound which would impart a suitable flavor to the finished product. Commercial butter-like flavors available today consist of various blends of diacetyl, acetyl methyl carbinol, butyric acid, ethyl butyrate, fatty acid lactones, ethyl vanillin, and similar compounds.

The choice of a particular blend of flavors depends on the use to which it is put. Blends which are highly acceptable in margarine may be objectionable in shortenings and oils. The presence or absence of an aqueous phase in the system has a profound effect on the odor and flavor characteristics of the product involved.

Diacetyl is highly volatile and disappears rapidly on heating of products containing it. Margarine usually loses its flavor by being held at room temperature for 1 or 2 days. Frying oils lose diacetyl long before they reach frying temperature. Butyric acid and the lactones are less volatile and are useful in products requiring heat stable flavors. Butyric acid, however, has a disagreeable odor when used alone. It is effectively modified and even masked by the presence of diacetyl. Therefore, butyric acid is never used without at least a small amount of diacetyl in the flavor compound.

BIBLIOGRAPHY

ANON. 1966. Versatile new baking ingredient prolongs shelf life and improves quality of yeast-leavened bakery products. Food Process.-Marketing 27, No. 5, 73.

ANON. 1968. Cuts filtration time after winterizing 50%, increases yield of winterized oil 8.4%, cold tests increased 30%. Food Process. 29, No. 12, 47.

BABAYAN, V. K. 1968A. The polyfunctional polyglycerols. Food Prod. Develop. 2, No. 2, 58, 60, 61, 64.

BABAYAN, V. K. 1968B. Esters of the polyglycerols. Food Prod. Develop. 2, No. 3, 84, 86, 90.

BABAYAN, V. K., and ATIKIAN, A. K. 1960. Sugar ester preparation. U.S. Pat. 2,948,717. Aug. 9.

CHANG, S. S., DE VORE, F. L., and FRIEDMAN, M. A. 1960. Shortening emulsifiers for use in icings. U.S. Pat. 2,966,410. Dec. 27.

FEUGE, R. O., ZERINGUE, H. J., WEISS, T. J., and BROWN, M. L. 1970. Preparation of sucrose esters by interesterification. J. Am. Oil Chemists' Soc. 47, 56–60.

GRETTIE, D. P. 1936. Salad oil and method of making same. U.S. Pat. 2,050,528. Aug. 11.

GRIFFIN, W. C. 1965. Emulsions. *In* Kirk-Othmer Encyclopedia of Chemical Technology, Vol. 8, 2nd Edition. John Wiley & Sons, New York.

HALL, L. A. 1957. Fatty monoglyceride citrate and antioxidant containing same. U.S. Pat. 2,813,032. Nov. 12.

HASS, H. B., SNELL, F. D., YORK, W. C., and OSIPOW, L. I. 1959. Process for producing sugar esters. U.S. Pat. 2,893,990. July 7.

KIDGER, D. R. 1962. Shortening emulsifier and method for preparing the same. U.S. Pat. 3,042,530. July 3.

KNIGHTLY, W. H., and KLIS, J. B. 1965. Seventeen ways to improve foods. Food Process. *26*, No. 5, 105–110, 114.

KUHRT, N. H., and BROXHOLM, R. A. 1962A. Method for preparing bakery products using mixed partial ester compositions. U.S. Pat. 3,034,897. May 15.

KUHRT, N. H., and BROXHOLM, R. A. 1962B. Mixed partial ester compositions. U.S. Pat. 3,034,898, May 15.

KUHRT, N. H., and BROXHOLM, R. A. 1963. Conjoined crystals. II. Applications. J. Am. Oil Chemists' Soc. *40*, 730–733.

KUHRT, N. H., BROXHOLM, R. A., and BLUM, W. P. 1963. Conjoined crystals. I. Composition and physical properties. J. Am. Oil Chemists' Soc. *40*, 725–730.

LITTLE, L. L. 1949. Mixed hydroxy and fatty acid glycerol ester emulsifiers. U.S. Pat. 2,480,332. Aug. 30.

MACDONALD, I. A. 1968. Emulsifiers: Processing and quality control. J. Am. Oil Chemists' Soc. *45*, 584A, 586A, 616A–617A, 619A.

MACDONALD, I. A., BLY, D. A., and SAMUEL, O. C. 1966. Short cut to maximum product quality. Food Process.-Marketing *27*, No. 5, 66–68.

MARTIN, J. B. 1953. Stabilization of fats and oils. U.S. Pat. 2,634,213. Apr. 7.

MELNICK, D. 1961. Flavor-stabilized salted margarine. U.S. Pat. 2,983,615. May 9.

MELNICK, D., and ACKERBOOM, J. 1961. Nonpourable soybean oil dressings. U.S. Pat. 2,983,618. May 9.

NASH, N. H., and KNIGHT, G. S. 1967. Polyfunctional quality improvers, polyglycerol esters. Food. Eng. *39*, No. 5, 79–82.

OSIPOW, L. I., and ROSENBLATT, W. 1967. Microemulsion process for the preparation of sucrose esters. J. Am. Oil Chemists' Soc. *44*, 307–309.

OSIPOW, L., SNELL, F. D., YORK, W. C., and FINCHER, A. 1956. Fatty acid esters of sucrose. Ind. Eng. Chem. *48*, 1459–1464.

POHLE, W. D. *et al.* 1964. A study of methods for evaluation of the stability of fats and shortenings. J. Am. Oil Chemists' Soc. *41*, 795–798.

RADLOVE, S. B., IVESON, H. T., and JULIAN, P. L. 1960. Oil soluble emulsifying agents. U.S. Pat. 2,957,932. Oct. 25.

SHAPIRO, S. H. 1961. Mixed glycerol esters of fatty and lactic acids for shortenings. U.S. Pat. 3,012,048. Dec. 5.

STAPF, R. J. 1959. Emulsified salad dressing. U.S. Pat. 2,885,292. May 5.

SWANSON, E. C. 1955. Performance testing. J. Am. Oil Chemists' Soc. *32*, 609–612.

TODD, P. H., JR. 1964. Vegetable base food coloring for oleomargarine and the like. U.S. Pat. 3,162,538. Dec. 22.

TRUAX, H. M, and MacDONALD, I. A. 1960. A determination of the effects of several variables on the performance characteristics of shortening using statistical experimental designs. J. Am. Oil Chemists' Soc. *37*, 651–657.

TUCKER, N. B. 1943. Surface active compound. U.S. Pat. 2,329,166. Sept. 7.

WEISS, T. J. 1967. Salad oil manufacture and control. J. Am. Oil Chemists' Soc. *44*, 146A, 148A, 186A, 197A.

WILKE, W. H. 1968. Newly approved emulsifier shows versatility in plant tests. Food Process. *29*, No. 5, 24–25.

WOODS, G. E. 1961. Edible emulsifying agents from glycerol and lactic and fatty acids. U.S. Pat. 3,012,047. Dec. 5.

YOUNG, H. H., and SPITZMUELLER, K. H. 1960. Shortening emulsifiers. U.S. Pat. 2,963,371. Dec. 6.

ZIEMBA, J. V. 1966. Today's monoglycerides do more for you. Food Eng. *38*, No. 1, 76–78, 80.

CHAPTER 5

Shortening—Introduction

Shortening once referred to the solid fatty material, such as lard, and its effect on making flaky and tender pastry crusts and crackers. The definition was later extended to include cake baking and frying fats. Recent promotion of unhydrogenated, polyunsaturated vegetable oils led to the phrase "liquid shortenings." All fats and oils are now called shortenings to distinguish them from margarines and other high fat content products which contain various nonfat materials in their composition.

SOLID SHORTENINGS

Solid fat products are the most generally useful and most varied of the shortening types. They have been formulated in many ways depending on raw material sources and end usage requirements. Shortening formulation can be deceiving in its complexity.

Solid shortenings are classified in one way according to plastic range. This defines the change in consistency of a chilled or plasticized shortening with change in temperature. A shortening with a narrow plastic range would be hard and brittle at low temperatures, e.g., in the refrigerator, and be soft and practically fluid at elevated room temperature. A product with a wide plastic range would be plastic and workable both in the refrigerator and at elevated room temperature.

Fully hydrogenated fats are used as plasticizers in shortening formulations. The word "plasticizer" describes its function in that the addition of hard fat extends the plastic range of the shortening. The widest plastic range possible is obtained by chilling a blend of liquid oil with nothing but hard fat. The amount of crystalline fat suspended in the oil base in such a blend remains practically unchanged from low temperature up to just under the melting point of the shortening.

High stability shortenings formulated from soybean oil hydrogenated under conditions which give high selectivity to about 70 iodine value and plasticized with about 5% hardfat would have a narrow plastic range. The typical all-purpose shortenings prepared from partially hydrogenated soybean oil at about 88 iodine value and 10 to 15% hardfat would have a moderately wide plastic range. The SFI values of these shortening types are given in Table 17. SFI values correlate with firmness of a shortening since consistency depends on the amount of solid fats in the shortening.

The addition of hardfats affects SFI values of shortening base oils more

81

TABLE 17

SFI VALUES FOR TYPICAL SHORTENINGS

Shortening	Plastic Range	Melting Point °F	SFI Value				
			50°F	70°F	80°F	92°F	100°F
High stability[1]	Narrow	109	44	28	22	11	5
All-purpose[1]	Wide	124	28	23	22	18	15
14% Hardfat in cottonseed oil[2]	Wide	124	16	14	14	12	11

[1] Weiss (1963).
[2] Mattil (1964).

at high temperatures than at low. A general rule is that SFI increases in value in an amount slightly greater than the percentage of added hardfat at 104°F, at an equal amount at 92°F, and at about 60% of the added amount at 50°F.

The correlation between amount of hardfat added to a shortening and the consistency of that shortening anticipates that the hardfat must have a beta-prime crystal structure. Beta crystalline fats would merely grow into larger granules if added beyond a minimum amount required for firmness. The effectiveness of the additional hardfat would thus be lost. Fully hydrogenated lard added to lard to be plasticized reaches a maximum consistency at about 10% hard lard. Further addition of hard lard not only would not bring about an increase in consistency, it might even cause the consistency to be reduced. Hard lard would promote the growth of crystals that might have remained more evenly distributed at lower concentrations.

Hydrogenation methods affect plastic range. Reduced selectivity results in a flatter SFI curve and consequently in a wider plastic range. The hydrogenation of margarine oils was usually carried out at the highest level of selectivity to obtain a sharp melting fat. Later demands for a margarine which was spreadable immediately when removed from the refrigerator called for a dual plastic range. The margarine would have to have a wide plastic range between refrigerator and room temperature, converting to a sharp melting product in the mouth. The solution to the problem was to blend oils hydrogenated at two different levels of hardness. Later developments have called for the blending of three components. Table 18 gives the SFI values for two component fats, a margarine oil blended from them to produce a spreadable margarine, and a single component margarine oil of the nonspreadable type.

Crystal structure is important in shortening compositions, the beta-prime phase being preferred for creamable cake and icing shortening. Where amounts of hardfat in excess of 5% of the formula are used to plasticize

TABLE 18

HYDROGENATED SOYBEAN OIL BLEND FOR MARGARINE

Oil	Iodine Value	Melting Point °F	SFI Value				
			50°F	70°F	80°F	92°F	100°F
Hard component	55	111	66	59	57	43	27
Soft component	86	86	18	7	2	0	0
20% Hard, 80% soft	80	100	27	16	12	3.5	0
Single component	75	100	32	18	11	3	0

Source: Mattil (1964).

the shortening, the crystal structure of the hardfat by itself is the factor which determines the crystal structure of the entire shortening. Fats which are chilled without hardfat must contain a minimum of 20% of a second component fat to ensure that the blend has a beta-prime crystal structure. This was previously done by blending at least 20% cottonseed oil with soybean oil before hydrogenation. It was then determined that a blend of 20% or more of a hard soybean oil (60 iodine value) with a softer soybean oil (80 iodine value) would be equally effective in maintaining a fine grained crystal structure (Merker *et al.* 1958).

If meat fats are permitted in the formulation, the minor component may be beef tallow or interesterified lard. These fats behave similarly to cottonseed oil. The major component should be a weak beta crystal fat, e.g., hydrogenated soybean oil. Fats which have strong beta crystalline tendencies, e.g., lard, must be restricted in use in a shortening which must crystallize in the beta-prime phase. The beta-prime component in this case must be added at a minimum of 55–60% of the formula to overcome all of the beta crystal forming tendencies of lard.

The addition of hardfats, emulsifiers, and antioxidants to solid shortenings is only a minor problem. These materials are soluble in hot oil. Once they are thoroughly dispersed, they will remain so as long as the shortening composition is held above its melting point. Chilling the shortening into its final solid form fixes these components into a permanently dispersed state.

Antifoam agents present a different problem. The silicones are not soluble in oil. They are also added in very small quantities, 0.5 to 1 oz per average shortening batch of 30,000 lb, and must be thoroughly dispersed. To complicate matters, there is evidence that silicones may deposit on metal surfaces, such as in holding tanks, pipes, pumps, and valves. Dispersion problems can be minimized by first dispersing the silicone in a small amount of oil or some volatile solvent such as toluene. The oil concentrate can be added to the shortening makeup tank just prior to chilling of the shortening. The solvent solution must be added to the deodorizer, however, in order to assure complete removal of the solvent. Care must be taken here to add the volatile solution to the deodorizer in such manner that the silicones are well dispersed in the finished oil and not merely deposited at the point of injection into the deodorizer. Chilling the solid shortening will, of course, maintain the silicones in whatever dispersion they had attained immediately before chilling.

Butter-type colors and flavors have been added to shortenings on occasion. Both of these materials are sensitive to heat. They must be added shortly before chilling and dispersed rapidly. Color in a solid shortening creates an additional problem in that its intensity is related not only to concentration of actual pigment but to gas (or air) content and bubble size. The most intense color would be obtained with a shortening containing no dispersed gases. The color decreases in intensity with increased whipping. A large volume of entrapped gas, broken into very fine bubbles, could result in a cream colored shortening in place of the desired butter color.

Pure white seems to be the most desired color for solid shortenings. A number of factors contribute to the visual color of a shortening. It should be recalled that whiteness depends on the amount of light reflected from a material. Light reflection in a shortening depends in part on the size distribution and number of gas bubbles and of fat crystals. The surfaces of these particles act as tiny mirrors dispersed in the shortening.

The color of the oils from which the shortening was formulated has a definite bearing on the whiteness of the finished shortening. Light striking the shortening not only reflects from the uppermost surface, but penetrates to some depth as well. Dark oils will yield off-white shortenings. As was pointed out previously, some oils are naturally darker than others, some may have been mishandled.

Storage of finished oils in melted form for subsequent shortening formulation causes progressive darkening. This also happens to finished, chilled shortenings. Oxidation of fats and fatty components, loss of gas, coarsening of solid fat crystals, all contribute to darkening of shortenings in storage.

Emulsified shortenings have color problems in addition to those shared by unemulsified products. Some emulsifiers, such as lecithin, are naturally dark. Iron oxides (rust) are soluble in monoglycerides and in many other emulsifiers. Older refineries use black iron equipment which can rust if the oil layer on the metal surface is washed away. Neutral oils apparently do not discolor in the presence of rust. Emulsified oils are prone to pick up rust and can become quite dark.

Packaging materials, especially plastic cube liners, can affect color perception. Shortenings are usually filled in cubes with blue liners which make the shortening appear whiter than normal. Occasionally a manufacturer will color code his shortenings by using different colored liners. Yellow, red, and green liners have been used and tend to alter the appearance of the shortening to make them seem less white.

Solid shortenings are subject to a number of texture and appearance defects. The most obvious is the occurrence of streaking. The Votator chilling machine has a spring loaded extrusion valve at the end of the system which is adjusted to give the desired back-pressure (250–400 psi). If there were no back-pressure to impede the shortening flow, it would be pumped directly through the system and would not be properly chilled. Streaking will occur if the shortening is chilled to too low a temperature for a given back-pressure. This is dependent on the relative hardness of the formulation. The action of the extrusion valve on a shortening which has been crystallized to an excessive degree is to literally squeeze the dispersed gases out of it. The shortening stream then consists of a mixture of aerated and deaerated product which appears as alternate streaks of white and cream color.

Too low a chilling temperature can cause other less obvious defects in solid shortenings. These include the presence of lumps of harder shortening in a softer matrix and shottiness which consists of smaller but harder lumps of about the size of buckshot. Occasionally the shortening seems to be in alternating layers of hard and soft product, similar to grain in wood. In a sense it is like streaking, but it is not visible since the gas distribution is not affected. These defects are observed by running the hand slowly across the shortening surface with the fingers digging a furrow an inch or two deep.

Oil separation is sometimes observed in shortenings which have a poor plastic structure. It will usually be found in little pockets of free oil. These pockets were originally air spaces. A partially emptied container of shortening will sometimes have a puddle of oil collect in a low spot. Oil leakage occurs if the shortening is stored at elevated room temperature but is less prone to occur if the crystal structure is of a definite beta-prime type. Even beta-prime shortenings will leak oil if they are formulated with an insufficient amount of hard fats.

FLUID SHORTENINGS

Fluid shortenings are a special type to be distinguished from liquid shortenings. Both are pourable but the liquids are clear while the fluids are opaque due to the presence of suspended solids.

The suspended solids may be hardfats or emulsifiers depending on whether the shortenings are to be used for frying or for baking (Cross and Griffin 1956; Schulman 1958; Brock 1959; Linteris 1959; Thompson 1959;

Eber *et al.* 1961; Houser 1961; Schroeder and Houser 1961; Howard and Koren 1965). Monoglycerides, lactylated monoglycerides, and propylene glycol monostearate are the most commonly used emulsifiers in fluid shortening formulations. The amount of hard material that may be added to the base oil would depend on the solubility of that material. The amount of solids in actual suspension should be 2–10%.

The size of the suspended particle is critical. Too large a particle would settle too fast and be too heavy to prevent crushing of the crystal. Too small a particle would settle very slowly but would settle eventually since small particles can pack closely together. This would rule out the use of beta-prime crystals as being too small. Beta crystal formers can be made to produce the right sized particles. They are prevented from settling by their physical dimension which is too large to pack closely but not so dense as to crush each other by their own weight.

Fluid shortenings can be prepared in several ways (Andre and Going 1957; Payne and Seybert 1961). One method is to cool the melted shortening slowly with slow agitation until it is properly crystallized. The process requires 3 to 4 days which is much too long to be commercially feasible.

A second method calls for grinding the hardfat or hard emulsifier to a fine powder. This is then suspended in the base oil without melting it. The coarse suspension is homogenized by passing it through a pressure homogenizer such as is used in dairy work. Colloid mills and shear pumps have also been used (Haighton and Mijnders 1968).

The method usually used for commercial production of fluid shortenings calls for rapid chilling of the product through a Votator A-unit or similar machine and holding it in a tank for at least 16 hr (Eber *et al.* 1961; Kearns 1962). The shortening will set up to a soft solid. It is then stirred gently to fluidize it and is filled into containers.

Correct temperature is important for storage of fluid shortenings. Storage below 65°F will cause the shortening to set up and lose fluidity. This can be reversed by warming. Storage over 95°F will result in partial or complete melting of the suspended solids. This is not a reversible phenomenon since cooling will cause formation of large crystals which will not remain in suspension. If the solids are hardfats, they will settle out in the container where their loss might be of little consequence. If the solids are emulsifiers, their settling out would upset the balance of the shortening formulation. The upper layer would be underemulsified, the bottom, overemulsified. Very little of the shortening would perform as expected.

An interesting thixotropic shortening has been described by Dobson (1967). This product is an opaque, white plastic material at rest, becoming pourable on being shaken. It is prepared by melting a mixture of 86%

hydrogenated soybean oil with a 107 iodine value, 10% fully hydrogenated soybean oil, and 4% mono- and diglycerides. The mixture is chilled to 60°F in 2 min and held for 30 min. It is heated to 105°F and held with agitation for 65–70 min. The mixture is then heated to 118°F, held for 20 min, cooled to 99°F, and pumped through a homogenizing valve at 250 psig. The heating and holding stages are designed to produce the appropriate crystal structure which would appear to be a finely dispersed beta crystalline form.

LIQUID SHORTENINGS

The liquid shortenings include regular cooking oils, salad oils, and the oils resulting from fractionation of semihard fats. They are not ordinarily emulsified. Most of the active emulsifiers are not sufficiently soluble in oils or low enough in melting point to remain clear at appropriate levels of use. If liquid shortenings lose clarity, they should be classified as "fluid." The main advantage of having a clear liquid emulsified shortening as opposed to a fluid type would be that the liquid product would not have any solid matter to settle out and thereby become altered in composition.

Production of the liquid shortenings was discussed in Chap. 3 under Winterization and Fractional Crystallization.

Ordinary cooking and salad oils offer no unusual problems in storage as they have no crystal structure or suspended solids to cause concern. If they are held at a sufficiently low temperature to cause solidification, they are easily melted by returning the oil to normal storage conditions.

Certain oils which have been prepared by fractionation of partially hydrogenated vegetable oils with an iodine value range of 70–75 can be the source of a peculiar storage problem. These are really quite hard. While they are liquid at low room temperature (70°F), they tend to become fairly solid at not too much lower a temperature (60°F). When reheated without agitation, they often do not remelt completely. The harder crystalline material tends to settle out and concentrate on the bottom of the container. Since melting is in fact the solution of harder fats in liquid oils, the concentration of solid fats on the container bottom now exceeds its solubility in the reduced amount of liquid oil available in that area. Apart from the presence of undesirable solid fats on the bottom of the container, these solids also tend to remove silicones as they settle out of suspension. It is advisable to mix the contents of containers of such liquid shortenings before using them in order to resuspend any settled solid fats and silicones.

POWDERED SHORTENINGS

Powdered shortenings are fats encapsulated in a water-soluble material. The fats are homogenized in a solution of nonfat solids. Emulsifiers may be added. The fat content of the powder ranges from 50 to 82% fat by weight.

Kraft (1936) used cheese whey solids as an encapsulating material. Fechner (1936A, B) used skim milk. North *et al.* (1947A, B) combined skim milk with corn syrup solids. They added monoglycerides, sorbitan monostearate, and lecithin as emulsifiers and alginate gums as a stabilizer. They used either spray or roller drying. Bogin and Feick (1951) encapsulated fat in methylcellulose or sodium carboxymethylcellulose.

Hansen and Linton-Smith (1966) have developed a more involved process for producing free-flowing powdered shortening. They claim that cake batters containing their product will cream readily. Their process calls for preparing a cream by homogenizing the fat in skim milk. The cream is mixed with a sodium caseinate-sodium citrate solution in water containing a dispersion of glycerol monostearate. The mixture is spray dried and the resulting powder is mixed with sodium silicoaluminate to make it free-flowing. It is claimed that the addition of monoglyceride to the cream after homogenization results in partial deemulsification of the fat during cake batter preparation. The partially released fat creams more easily than the better dispersed fats of other powdered shortenings.

PACKAGING

Proper packaging involves the correct size and shape of the container and the materials from which the container is made. Most people are familiar with grocery items and the bottles, cans, and boxes used to hold them. They are designed to attract the customer as well as to protect the contents of the package. Convenience of use is also a factor but may be considered as part of the attractiveness of the container. Attraction is often considered to be more important than protection where the two conflict. The argument here is that high turnover rate offsets the need for long shelf-life protection.

Packaging of household shortenings and of other products will be discussed in later chapters dealing with those products. Consideration will be given here only to bulk packaging for industrial and commercial shortenings.

Attractiveness of the bulk container is not ignored but is secondary to cost, protective ability, or convenience.

Solid shortenings are usually filled in 50-lb cans and cubes, 110-lb cans, and open-end steel drums holding about 380 lb shortening. The shortenings are plasticized to make them uniform throughout their mass. Any one portion of the shortening should be the same as any other portion in the same container. Plasticizing also makes the shortening more attractive in appearance. It is also essential for the creamability of solid shortenings.

The friction top can offers a convenience to the person who uses only part of the contents of the container at a time. It is easy to reclose and store. The partially emptied can will stack without crushing since it is rigid and self-supporting. In the case of frying fats, the melted shortening can be poured back into the container if desired. This is especially convenient for filtering the fat and while cleaning the fryer. The empty cans are also reusable and are often used for storage of various assorted items about the kitchen or bakeshop.

Cubes are corrugated paper boxes measuring about 1 ft on a side. Some shortening suppliers use other dimensions but their volume approximates 1 cu ft. The boxes are lined with a bag of polyethylene film to keep the shortening from leaking through the box. This is the least costly of the small package containers. It is most convenient for the user who would use the entire contents of a carton at one time. The polyethylene bag of shortening is easy to remove from the box. The bag can then be stripped off, exposing the free-standing cube of shortening. The empty boxes and bags can be folded flat for disposal. In this way, they take up a minimum of space until they are hauled away.

The plastic cube liners are usually highly colored. They melt in hot fat and may accidentally fall into a fry kettle. The color makes them easy to detect and to retrieve from the fryer. Melted cube liners have been known to cause foaming, darkening, and smoking of the frying fat.

Drums represent the lowest cost packaging for solid shortenings as they are usually resold to be cleaned and refilled. Bakery shortenings which are used in solid form are somewhat difficult to remove from a drum as they must be scooped out by hand. Frying fats or other shortenings which can be melted for use are easier to handle. There are a number of drum melters available on the market. They consist of immersion heaters and heated jackets to be placed around the drum. They may be heated electrically or by steam. It is also possible to maintain a small cabinet or hot room to melt and hold several drums at one time.

When a drum of shortening is melted from a plasticized product, the liquid fat will be uniform throughout. If the drum is filled from a tank of melted fat or if previously melted plasticized fat is allowed to solidify, the remelted product will not be uniform. It must be stirred thoroughly if a portion of it is to be removed. This is because the harder fats in a shortening solidify first on cooling and, being more dense than the liquid portions, settle to the bottom of the drum. They will remain there, even on complete melting, so that the shortening will be stratified into layers of unequal hardness.

Fluid and liquid shortenings are usually packed in 1, 2, and 5 gal. cans and closed-end drums holding 55 gal. There are no special problems attached to the physical handling of pourable shortenings or oils. The container size is selected by the requirements of the user. For example, a 5-gal. can of oil weighs about 38 lb and is difficult for a woman to handle. One- and even 2-gal. cans are preferred in kitchens where women cooks are employed.

Drums of pourable product may be emptied conveniently in 1 of 2 ways. The drum may be held upright and fitted with a pump, usually hand-operated, or it may be fitted with a spigot, put in a special rack in which the drum can be tipped on its side and emptied by gravity.

Fully hydrogenated fats and fatty compounds such as hard monoglycerides are usually flaked by pouring the melted fat in a thin layer onto a revolving chilled drum and scraping the hardened material from the drum. They are also chilled by spraying them into a blast of cold air in the large chamber of equipment similar to that used for the spray drying of milk powder. The flaked or powdered material is then packaged in multi-walled paper bags or fiber drums. The bags are usually 50-lb units while the drums hold 100–200 lb.

Confectioners' hard butters may be flaked or powdered as they are sufficiently hard at room temperature to be handled in this manner. They are also available in 5- or 10-lb cakes or slabs. They are made by pouring the melted fat into pans or molds where they are permitted to stand until they harden and can be dumped by hand. Several slabs are then placed in a corrugated paper box. The slabs are separated by sheets of paper to prevent them from sticking to each other.

BULK HANDLING

Manufacturers who use solid shortenings in sufficiently large quantity, in the order of at least 20,000 lb per week, most frequently purchase their shortening supplies in bulk. Shipments are made by 20,000-lb capacity tank wagons or 30, 60, or 150,000-lb capacity tank cars. There are a number of factors to be considered in bulk handling installations.

The presence of a railroad siding should be one of the first considerations. Without a siding, the shortening user has to rely on tank wagon deliveries and is, therefore, tied to a supplier within about 500 miles of his plant. Bulk deliveries must be in a liquid state to be removed from the carrier. Tank wagons usually have no facilities for remelting fats. The shortening must be delivered and emptied before it solidifies. An insulated tank filled with a hot shortening will remain fluid for about 12 hr. The length of time depends on filling temperature and melting point of the oil, ambient temperature, and other weather conditions. The type of highway and traffic conditions in general help determine how far a tank wagon can be sent before the shortening it contains begins to set up. A miscalculation can cause many headaches in trying to remove a hardened shortening from the tank.

Tank cars are usually provided with steam coils for remelting shortening. Some modern cars are equipped with heaters which maintain the shortening in a molten state until delivered. Manufacturers with limited permanent storage facilities have used tank cars for temporary storage even though demurrage charges are not low. The savings involved in bulk handling can offset the cost of an occasional storage problem.

On receipt, the melted shortening is pumped into storage tanks to be held until needed. The shortening should be maintained at a temperature 5° to 10°F above its FAC melting point. It is also advisable to blanket the shortening in the tank with nitrogen gas. The shortening will deteriorate slowly in storage and become darker in color. The low-storage temperature and nitrogen blanketing help retard this deterioration, but the storage life of melted shortening is limited. The tank of shortening should be turned over in 7 to 10 days for best results. The capacity of the storage tank should be based on this turnover rate.

Storage tanks are fabricated in many shapes, round or square, conic or flat-bottomed, tall or short. The design selected depends on space requirements which are often critical in a modern plant layout.

Tanks may be heated by a number of methods. Steam coils are inexpensive but are subject to leakage and can cause local overheating of fats if proper care is not taken. Some tanks are jacketed. The most frequently used method, however, is the hot room. The storage tanks are merely held in a room which is maintained at the desired temperature.

Storage tanks are rarely equipped with mechanical agitation. Some mixing occurs when new shortening is pumped into a partially emptied tank. The ability to agitate shortening in storage is sometimes desirable, however. Fats that harden in transit and are remelted on delivery may get sufficient mixing on being pumped into storage, but this is not always the case. This is especially important for emulsified shortenings. Failure to maintain proper storage temperature causes the harder fats to solidify and settle out. Agitation is required to remix these fats on melting. If mechanical mixers are not available, the shortening can be agitated by blowing nitrogen through the tank outlet, allowing the bubbling action to mix the oils. The oils can also be mixed by circulating them with a pump connected between tank outlet and a line to the top of the tank. This latter method should be done carefully so as not to aerate the melted shortening excessively.

Bulk handling systems for bakery shortenings require equipment in addition to storage facilities, specifically equipment for chilling and plasticizing the shortening. Freshly plasticized shortening may be added to dry cake mix without previous tempering as it seems to temper in the finished mix.

PERFORMANCE TESTING

There are some necessary or desirable attributes of shortenings which are not measurable by chemical or physical tests. Performance tests are the only means of evaluation in such a case. This calls for setting up a procedure which simulates actual usage conditions. It is not always easy as many plant processes do not scale down properly. Testing procedures must be modified and carefully correlated with full scale plant operations.

Performance testing is essential in the development of new products. It is the only way to evaluate the effectiveness of a new emulsifier or a new use for an old emulsifier. The same would be true for other additives, such as antifoam agents, antioxidants and sequestrants. Care must be taken in interpreting results, in not placing too much emphasis on laboratory performance as the ultimate when operating performance is the true goal.

Bakery products are tested by the actual baking of cakes, creaming of icings, etc. The question, however, is, "what cake, what icing?" Shortening manufacturers, emulsifier suppliers, and others have developed their own test formulations and their own test procedures (Erickson 1967). They have then proceeded to improve their products to give the best results with their own tests. Competitive surveys often reveal that the manufacturer making the survey gets the best results from his own product.

The shortening user also has his special tests. If the supplier's product passes his test, the product is added to the list of approved shortenings. Unfortunately, the supplier does not always have access to the purchaser's test methods.

Most performance tests are designed to be hypercritical. They are to help in developing improved shortening formulations and to sort out occasionally bad batches of product. Commercial bakery products are designed to be tolerant of raw material variation and to make acceptable finished baked goods under most conditions. This tolerance reduces the responsibility of performance tests in evaluating a shortening. The shortening would have to be badly misformulated or mishandled to fail in commercial use.

Frying performance tests are often run by frying small amounts of food at regular intervals in the product under test until the fat breaks down. Frequently the test is run with potatoes which contribute very little suspended or dissolved matter to the shortening. Actual restaurant practice calls for frying an assortment of foods. Chicken, fish, egg batters, breadings, doughnut batter, assorted foods, all contribute suspended solids and, in many cases, highly unsaturated oils to the frying shortening. It is not too long after frying begins that the frying fat does not resemble the original shortening in composition.

Some tests on frying shortenings have been performed by heating the shortening until it breaks down but without actually frying in the shortening. Such tests have no relation to reality since the moisture released during frying removes various fatty breakdown products, preventing them from accumulating in the fat. Fats heated without frying soon become chemically different from those heated with frying. This difference can invalidate the test.

Determination of the flavor and odor of oils is in a sense a performance test. Blandness is the goal for most oils and shortenings. Care must be taken in evaluating emulsified products. Monoglycerides, for example, contribute a feeling in the mouth which, while bland in flavor, may not be interpreted in this way.

Blandness cannot always be the desired goal. The flavor of lard should be meaty. Some types of lard taste normal only if they are slightly rancid. The flavor of virgin olive oil is highly prized, as is the delicate, perfumey odor of cocoa butter. These flavors and odors cannot be detected by chemical means.

SPECIFICATION WRITING

Specifications are required in all phases of fat and oil processing and usage. This is true for production of all materials, but experience has shown that special considerations required for fats and oils are not always well understood. The sections in Chap. 1 concerning methods of evaluation describe some of the problems encountered in the interpretation of results obtained. These methods are the basis for the specifications.

Specifications are the guidelines of the shortening processor for the purchase of raw materials, processing of intermediates, and formulation of the final product. This last item becomes a raw material for the shortening user who then develops his own specification.

It is important for the sake of simplicity and economy to incorporate only the essential requirements into a specification. Superfluous analyses are costly to determine and may even be ignored. Omission of key information could be disastrous but this situation is rare.

Crude oils are purchased on the basis of refining loss and refined and bleached color. Very little else is required. Oils in process are checked for hardness if they are hydrogenated, using melting point, iodine value, and SFI at certain critical temperatures depending on the product. Salad oils are checked for cold test and color.

Many finished shortenings are sold under brand names. Specifications for them are developed by the producer. The average consumer has little knowledge of them since his primary concern is satisfactory performance. He looks mostly for an apparent uniformity. He is unaware of physical and chemical evaluations.

Producer specifications include various attributes, frequently SFI, melting point, penetration or consistency values at some temperature, monoglyceride content, free fatty acid, peroxide value, and color. The quality assurance departments use these specifications to determine uniformity, rejecting for shipment any product which fails to meet specification.

The bulk user of shortening products is well aware of specifications and is often equipped with an analytical laboratory. This has led to many an argument. Fats and oils are natural products. Their analysis is subject to more error than is found with pure chemical compounds. Check samples are used frequently to determine the ability of various laboratories to agree with each other on their results. Official referees are available to settle disputes.

In writing a specification, it is important to allow for analytical error as well as for reproducibility allowance in processing. A workable range of values should be assigned to each analytical criterion. This sometimes results in setting a maximum or minimum value rather than upper and lower limits.

All uses for shortenings have certain key requirements. This is often merely a melting point, emulsifier level, or textural attribute. The potential customer, however, has received a sample of a product which performs well. He has it completely analyzed and may use the complete analysis as a basis for his purchase specification. The description becomes quite involved and eliminates many potentially acceptable competitive shortenings which do not meet "specs" in all ways. This could penalize both supplier and user. An ironic situation sometimes develops when a supplier carelessly submits a shortening sample that is at the upper or lower limit of his normal range of production capability. The prospective customer sets his specification limits based on this sample. The original supplier then finds that only half of his production can meet the purchase specifications.

Quality specifications should always be included with those relating to physical character and performance. These include color, flavor and odor, moisture content, peroxide value, and free fatty acid level. Quality specifications insure careful handling, from good deodorization and filtration, through clean holding tanks, filling lines, and carrier tanks to adequate protection during delivery. This would also apply to plasticized shortenings which could deteriorate with long or improper storage.

Surprisingly, 2- or 3-yr old packaged shortening has been found on occasion after warehouse cleanup. The product is usually dark in color and is reduced in volume due to loss of entrapped air. Shortenings have also been stacked too close to heaters or steam pipes with consequent deterioration from excessive heat.

To summarize, purchase specifications should be considered with care to include only what is necessary. Anything beyond this can only lead to waste.

BIBLIOGRAPHY

ANDRE, J. R., and GOING, L. H. 1957. Liquid shortening. U.S. Pat. 2,815,286, Dec. 3.

BOGIN, H. H., and FEICK, R. D. 1951. Stable fatty-food compositions. U.S. Pat. 2,555,467. June 5.

BROCK, F. H. 1959. Liquid shortening. U.S. Pat. 2,868,652. Jan. 13.

CROSS, S. T, and GRIFFIN, W. C. 1956. Vegetable oil shortenings. U.S. Pat. 2,746,868. May 22.

DOBSON, R. D. 1967. Thixotropic shortening. U.S. Pat. 3,360,376. Dec. 26.

EBER, F., OTT, M. L., COCHRAN, W. M., and KEAN, H. J. 1961. Fluid shortening and process for making the same. U.S. Pat. 3,011,896. Dec. 5.

ERICKSON, D. R. 1967. Finished product testing. J. Am. Oil Chemists' Soc. 44, 534A, 536A, 538A, 561A.

FECHNER, E. J. 1936A. Comminuted dry shortening for use in making bakery products. U.S. Pat. 2,065,675. Dec. 29.

FECHNER, E. J. 1936B. Comminuted dry shortening for use in making bakery products. U.S. Pat. 2,065,675. Dec. 29.

HAIGHTON, A. J., and MIJNDERS, A. 1968. Preparation of liquid shortening. U.S. Pat. 3,395,023. July 30.

HANDSCHUMAKER, E., and HOYER, H. G. 1961. Fluid shortening and method of making same. U.S. Pat. 2,999,755. Sept. 12.

HANSEN, P. M. T., and LINTON-SMITH, L. 1966. Powder produced from butter or other edible fats. U.S. Pat. 3,271,165. Sept. 6.

HOUSER, C. J. 1961. Liquid shortening. U.S. Pat. 2,968,562. Jan. 17.

HOWARD, N. B., and KOREN, P. M. 1965. Fluid shortening for cream icings. U.S. Pat. 3,208,857. Sept. 28.

KEARNS, J. J., Jr. 1962. Fluid shortening. U.S. Pat. 3,028,244. Apr. 3.

KRAFT, G. H. 1936. Comminuted shortening. U.S. Pat. 2,035,899. Mar. 31.

LINTERIS, L. L. 1959. Fluid shortening U.S. Pat. 2,909,432. Oct. 20.

MATTIL, K. F. 1964. Bakery products and confections. In Bailey's Industrial Oil and Fat Products, 3rd Edition, D. Swern (Editor). Interscience Publishers Div., John Wiley & Sons, New York.

MERKER, D. R., BROWN, L. C., and WIEDERMANN, L. H. 1958. The relationship of polymorphism to the texture of margarine containing soybean and cottonseed oils. J. Am. Oil Chemists' Soc. 35, 130–133.

NORTH, G. C., ALTON, A. J., and LITTLE, L. 1947A. Shortening. U.S. Pat. 2,431,497. Oct. 25.

NORTH, G. C., ALTON, A. J., and LITTLE, L. 1947B. Shortening. U.S. Pat. 2,431,498. Oct. 25.

PAYNE, E. T., and SEYBERT, R. A. 1961. Liquid shortening method. U.S. Pat. 2,999,022. Sept. 5.

SCHROEDER, W. F., and HOUSER, C. J. 1961. Liquid shortening. U.S. Pat. 2,968,564. Jan. 17.

SCHULMAN, G. 1958. Liquid shortenings. U.S. Pat. 2,864,703. Dec. 16.

THOMPSON, S. W. 1959. Fluid shortening compositions. U.S. Pat. 2,875,065. Feb. 24.

WEISS, T. J. 1963. Fats and Oils. In Food Processing Operations, Vol. 2, M. A. Joslyn, and J. L. Heid (Editors). Avi Publishing Co., Westport, Conn.

CHAPTER 6

Bakery Shortenings

INTRODUCTION

It is a revelation to consider the myriad items that are produced in bakeries of all types. Practically all bakery products require the use of shortenings in one form or another. Many require shortenings in different component parts, e.g., a filled cupcake topped with an icing could require shortening for the cake, the icing, and the filler. Small bakeshops would use one all-purpose shortening for each of these functions. Large establishments could easily use three different fat products, each designed to be most efficient for the required end product.

Some highly specialized baked goods, such as Danish and puff pastry, use margarines instead of shortenings. In one respect, these are really shortenings in margarine form, i.e., fats emulsified with milk or water. Since they are manufactured as margarines, they will be discussed as such in Chap. 9.

Each shortening producer has his own set of formulas and his own operating methods. However, it is difficult to be different for too long a time. A successful shortening can be analyzed, imitated, and appear competitively on the market in about six months. Since actual formulations of commercial shortenings are highly secretive and subject to change with fluctuations of the oil market, any discussion must be somewhat general. Specific details when given are as accurate as the subject will allow.

ALL-PURPOSE SHORTENINGS

There are actually two general purpose shortenings being marketed, an emulsified and an unemulsified type. A bakery with nothing but each type on hand could produce a large variety of baked goods. The small local bakery was operated in this way. The larger city-wide bakeries which have supplanted the independent shop often rely heavily on all-purpose shortenings for the bulk of their output with a few specialized products for their own specialty items.

Unemulsified Shortenings

The basic general purpose shortening has a wide plastic range. It is usually formulated by adding 10–15% of a fully hydrogenated beta-prime hardfat to a partially hydrogenated vegetable oil, an interesterified lard, or a blend of each. Unhydrogenated tallow is also frequently used. The hardfat may be prepared from cottonseed oil or tallow. The vegetable oil base is usually hardened soybean oil as this is the lowest cost oil as a general rule. The base is normally hydrogenated to an iodine value range of 80 to 85.

All-vegetable shortenings are usually not stabilized. Antioxidants must be added to the meat fat types, however, as they are rarely hydrogenated and must be protected from oxidation.

An all-purpose shortening is a compromise between having the best frying fat and the best pastry and cookie shortening. It is also widely used for certain pound cake formulations. Further hydrogenation would make a more stable frying fat, but such a fat would not cream well or handle as well in making pastry. All-purpose shortenings may have a small amount of silicone added to them to improve frying life. The amount is less than the optimum for maximum frying life but more silicones would be detrimental to the baking of certain types of cookies and cakes.

While few bakeshops do much total frying, a sufficient number of them have a doughnut operation to warrant the large-scale manufacture of all-purpose shortenings.

Emulsified Shortenings

The emulsified general purpose shortening is formulated in the same manner as the unemulsified but with the addition of monoglycerides. The emulsifier may be a mixture of mono- and diglycerides or distilled monoglyceride. The level used will depend on the manufacturer but will usually be between 2.25 and 2.75% by analysis.

Silicones are never added to this type of shortening. Creaming ability is the most important function of this shortening. Silicones would impair and even destroy the incorporation of air and water into batters and icings. On the other hand, the emulsifiers which impart creaming ability reduce the smoke point of fat to the extent that it cannot be used for frying.

All-purpose emulsified shortenings represent a compromise between making the best cakes and making the best icings. Cake volume and structure are most desirable when the monoglyceride component is prepared from fully saturated fats. Icings are the smoothest and fluffiest when the monoglycerides are prepared from unhydrogenated oils. The compromise is reached by using monoglycerides made from partially hydrogenated oils with an iodine value range of 72–76 for soybean oil or 74–78 for cottonseed oil.

Much work has been done on evaluating the relative merits of cottonseed and soybean oil monoglycerides. The higher palmitic acid content of cottonseed oil seems to increase the effectiveness of monoglycerides when used at the same level as those prepared from soybean oil. A 0.5% additional soybean mono-diglyceride (40–42% mono content) in a shortening could offset the difference in performance. The cost differential between the two could be the deciding factor on which source oil is used.

A continuing search is being made for an emulsifier system which will make better cakes and icings at the same time. Thus far, most available emulsifiers which improve cakes degrade icings and vice versa. Some potent emulsifiers which work well for both have a bitter flavor that is especially noticeable in icings.

Beta-prime crystal structure is vital to the satisfactory creaming performance of emulsified shortenings. A shortening that is borderline between beta-prime and beta may deteriorate in storage and lose its creaming power.

A wide plastic range is important in general bakery work. Storage and shop temperatures are quite variable through the year between different shops and in various parts of the country. Wide plastic range shortenings are tolerant to broad variations in working temperature. They are also resistant to breakdown during creaming as they are soft and pliant. Lumpiness, shottiness, brittleness, and other textural defects interfere with creaming and can result in uneven cakes and icings.

SPECIALTY BAKING SHORTENINGS

Cake Shortenings

Many specialized shortenings have been developed for cake baking. Each is designed to give the best results for a given cake formulation. In some cases, special cakes have been developed to take advantage of a new shortening formulation.

Standard Cakes.—Ordinary bakery cakes are sold over the counter, through home delivery, or wrapped in cellophane and sold through distributors to appear on grocery shelves a long distance from the bakeshop.

They may be iced mechanically or by hand. In any case, they are mass produced and must be rigid enough to hold up under relatively rough treatment. They must not become stale too readily.

Most standard cakes are baked with all-purpose emulsified shortenings. The baker might find that a special cake shortening made with a harder monoglyceride would improve his cake. However, the shortening manufacturer also benefits from all-purpose shortenings. It helps reduce the number of products he must stock. Only the largest baking companies with national distribution purchase in sufficient volume to obtain specialized shortenings for their cakes.

Since large companies maintain their own laboratories and test bakeries, they are in a position to follow the development of specialized shortenings. Most frequently, the shortening developed for the standard cake of a particular bakery is a regular all-purpose shortening with a slightly higher monoglyceride level. The shortening then fits the particular cake formula more closely, allowing for increased water in the formulation or a larger cake for a given weight of batter. Occasionally the hardfat content of the shortening is changed so that the hardness of the shortening is better adjusted to the specific needs of the bakery involved.

A number of more or less successful attempts have been made to market a fluid shortening for cake baking. The advantage obtained would be in ease of handling and measuring the shortening. This would be an especial advantage for small bakeshops in preparing small batches of batter. The emulsifiers used in such shortenings were the lactylated monoglycerides and propylene glycol monostearate (Schwain 1965).

The main difficulty in obtaining acceptance for such shortening was that new formulas and new techniques had to be sold to the baker along with the new shortening. In addition, fluid shortenings seemed to be less tolerant of normally variable bakeshop conditions than were ordinary solid shortenings.

Dry Mix Cakes.—Cakes prepared by the housewife are quite different from commercially baked cakes. This is especially true for those made from prepared mixes. Such cakes are supposed to be as large as possible from a given weight of mix. They are to be moist and tender. Hopefully, they will come out right no matter what the housewife does in mixing the batter and baking it.

The commercial baker would be happy to produce this type of cake if it would hold together. The housewife makes one cake at a time and handles it delicately. The baker cannot afford to. The dry mix type cake would crumble while being iced or packaged. If it reached the customer without disintegrating, it would probably be stale, as such cakes do not keep well.

Shortenings for cake mixes are highly specialized and highly individualized. Each mix formulator has shortenings designed for his own mixes. Each mix type prepared by one manufacturer often requires a different shortening than do the other mixes. Yellow cake using eggs in the batter is the basic mix. White cake, containing no egg yolk, requires modification of the formula to compensate for the lack of emulsification. Devil's food or chocolate cakes contain cocoa which affects the cake batter system. Shortenings may have to be modified to compensate for this.

The most important factor in prepared cake mix shortenings is the emulsifier system. The first emulsifiers used were ordinary hard mono-glycerides. Rapeseed monoglycerides were considered to be especially effective (Bedenk 1962). Many of the emulsifiers developed since then were designed for the cake mix industry. The lactylated monoglycerides, primarily lactylated monopalmitin, were among the first of these. Pro-pylene glycol monostearate (PGMS) came next.

PGMS was first used as the sole emulsifier at a level of about 20% of the shortening. The cake mix using PGMS would not tolerate a high fat level. Cake baked from it was too dry. It was soon learned that PGMS would make highly satisfactory cakes with normally high shortening levels when used in combination with hard monoglycerides.

The improvement in functionality by the combination of monoglycerides with other emulsifiers seems to be the general rule. Monoglycerides form an essential part of emulsifier systems for cake mixes and for other uses. The other component of the system may be one of a number of com-pounds, e.g., lactylated monoglycerides, polysorbate 60, sucrose esters, etc. Emulsifier suppliers list such combinations as standard items.

Lecithin is added to some shortenings at a level of 2–4% of the product. Such shortenings are designed specifically for use in white cake mixes to compensate for the absence of egg yolk. The addition of lecithin is not necessary and may be undesirable for other types of cake mix.

Plastic range is not important for prepared mix shortenings. The mixes are made under controlled conditions so that the shortenings are not re-quired to remain workable through a broad temperature range. Nor-mally, hydrogenated vegetable oil shortenings for mix use have a narrow plastic range as they are hardened further for oxidative stability. Meat fat shortenings would have a wide plastic range since those fats are nat-urally hard and have an inherent wide plastic range.

As with all cake shortenings, those for mix use must have a beta-prime crystal structure.

Some of the larger mix manufacturers purchase emulsified shortenings in bulk and chill them from storage in melted form. The economies ob-tained in bulk handling more than pay for cost of the chilling equipment.

Chiffon Cakes.—These cakes are made with unemulsified liquid salad oil as the only shortening.

A special type of plasticized cake shortening containing a large quantity of liquid oil appeared on the market. It produced a moist cake with the open texture of a chiffon cake but without its lightness. The shortening and its resultant cake did not achieve much popularity.

Liquid oil is difficult to use in cake baking. It is not tolerant to mis-handling. Temperature control is more critical with liquid oil batters than with batters made from solid shortenings. Various emulsifier systems have been tested in conjunction with liquid oil cakes but have not been able to overcome the difficulties involved in their use.

Yeast-raised Dough

Common white bread and sweet doughs used for Danish pastry use special shortenings which are primarily concentrations of monoglycerides. The U.S. FDA Standards for bread set the maximum monoglyceride content of the shortening at 20%. They were referring to the 40% concentrate which would result in about 8% monoglyceride content by analysis. When distilled (90%) monoglycerides are used, they are permitted at 10% of the shortening resulting in a 7–8% level by analysis.

Partially or fully hydrogenated lard is the usual source oil for preparation of monoglycerides for bread use. The main function of the monoglyceride is to soften and to retard staling of the baked product. The saturated monoglycerides seem to be more effective in retarding staling than the less saturated products. Some other emulsifiers seem to be active as antistaling agents. These are sodium stearyl fumarate and calcium stearyl-2-lactylate. They are permitted for use in bread by U.S. FDA Standards at 0.5 parts emulsifier per 100 parts of flour by weight.

The base or carrier fat for bread emulsifiers or softeners is usually the lowest cost fat available. This is usually lard or lard and tallow blends, where meat fats are acceptable, or hydrogenated soybean oil for the all-vegetable product. The deciding factor is cost as the base has no activity in retarding staling or any other effect on the finished baked product. There is a limit in the amount of tallow that can be used as both tallow and the monoglycerides are hard. An excessive amount of tallow would make the finished shortening too firm to handle easily.

At least one product of this type is colored yellow and flavored with a butter-like flavor.

It is also possible to use monoglycerides without a shortening carrier. Hard monoglycerides will disperse in hot water to form a creamy paste. This can be added directly to the dough. Softer 40% monoglycerides have almost 60% di- and triglycerides to act as a carrier. The diglycerides apparently have no antistaling activity on bread but would help disperse soft monoglycerides. It would not be practical to add the hard monoglycerides as such to the dough as they would be almost impossible to disperse properly.

Continuous Bread.—Shortenings for continuous bread production are somewhat similar to those made for regular bread baking. The main difference in usage is that regular bread shortenings are added to dough in solidified form while continuous bread systems rely on pumping a melted shortening into the product stream. The shortening must solidify quickly since the mixing and proofing time is so short.

Early development of continuous bread machines was done with full scale equipment. The first shortenings which seemed to work well were adopted immediately. Change was difficult to make as failure of a test could result in the loss of hundreds of loaves of bread. Pilot units were introduced later, as meaningful scaledown of test apparatus proved to be difficult.

The first successful shortening consisted of a harder than normal lard to which the baker added 10–40% fully hardened cottonseed oil flakes depending on his own requirements (Parker 1965). Monoglycerides were not incorporated at first. Further study led to a balance between hardfats and hard monoglycerides in the shortening.

Baldwin et al. (1965) report that the optimum composition for a continuous bread shortening consists of a blend of 4% hardfat flakes and 2% hard distilled monoglyceride suspended in 94% lard or soybean oil.

Pie Crust

The standard shortening for flaky pie crust has always been lard. The lard may be in its natural state or be deodorized and stabilized with antioxidants. Fully hydrogenated lard is often added to aid in plasticizing the product.

It has been postulated that the grainy crystal structure of lard formed a dough structure that contained many tiny layers of fat interspersed between layers of dough. These fat particles would melt on baking, leaving a flaky, tender crust. Smoother shortenings would supposedly work into the dough more thoroughly and not leave those pockets of shortening to separate the dough into flaky layers.

Cake-type pie crusts were not supposed to be flaky. They were prepared from all-purpose shortenings. The cake-type crust could be prepared by machine mixing and were popular with large bakeries. Flaky crusts were thought to require hand preparation to be satisfactory. Only small specialty bakers could afford to make flaky crusts.

Mechanical mixing methods were developed for preparation of flaky pie crust. The various bakeries developed their own techniques which required specific shortening attributes depending on the individual needs of the shop.

Proper mixing action depends on the firmness of the shortening which must not become too soft, and especially not melt during the operation. Hand mixing has been carried out using ice in the dough to keep the lard sufficiently firm.

Lard compounds for modern pie crust production are firmed up by the addition of varying amounts of fully hydrogenated lard flakes or lard stearine (if and when available). The overall firmness of the shortening would depend on the temperature at which the particular shop in question operated. A pastry dough to be mixed at a warm temperature, for example 70°F, would require a harder product than one to be mixed at 50°F. The shortening in each case would have an identical firmness at its particular mixing temperature.

One frozen pie producer required a lard to be workable at so low a temperature that the shortening had to be especially chilled to form large, grainy crystals having little plasticizing effect. The hard triglycerides were crystallized into a relatively small number of particles by under-chilling the lard. The shortening was filled into cubes and stacked in a cold room in such manner that the heat of crystallization was rapidly dissipated. The crystals were allowed to grow slowly for several days. The finished product was then shipped in refrigerated trucks to the customer. Needless to say, the pie maker soon modified his method. The cost of producing the specialized lard was reflected in his costs, making it difficult for him to be competitive.

All-purpose, unemulsified shortenings are used for pie crust preparation where an all-vegetable shortening is desired. Fluid shortenings and liquid oils have also been used, especially where nutritional claims are made referring to the polyunsaturated fatty acid composition of the pastries involved.

Crackers

Crackers are produced with high stability shortenings, specifically hydrogenated lard, hydrogenated vegetable oil, hydrogenated blends of tallow and soybean oil, and sometimes general purpose unemulsified shortening.

In addition to the shortening incorporated in the dough, some crackers are sprayed with a thin layer of coconut oil to act as a moisture barrier. This helps retain crispness in the cracker. Some prepared dry breakfast cereals are similarly treated for the same reason. The cereal remains crisp for a longer time when cold milk is poured on it.

Cookies

There are many types of cookies, soft and hard. The soft cookie requires creaming which is adequately furnished by the unemulsified all-purpose shortening. Hard cookies merely require lubrication. Any fat or oil could provide this.

Cost is a major factor in shortening selection for cookies. Deodorized and stabilized lard, hydrogenated blends of tallow and soybean oil, and hardened soybean oil are current examples of the lowest cost shortening types. The fat and oil market is always subject to change, however.

Cookie shortenings must be smooth and free from shot or lumps. Unevenness in shortening texture can result in pockets of pure shortening in the mixed dough. On baking, the shortening will melt and leak out leaving voids and uneven areas in the finished cookie.

Some types of cookies are enrobed with confectioners' coatings. This will be discussed as a separate subject in Chap. 12.

Pan Grease

Ordinary white bread is baked in loaf pans arranged in racks. The racks are inverted on leaving the oven. It is essential that the loaves release readily from the pans.

A number of pan greases have been developed to aid in release of the loaves. The grease must be easy to apply to the pan. This is ordinarily done by spraying and requires that the grease be a pumpable fluid.

Lard oil was once one of the most popular of pan greases as it had a natural lard flavor which was imparted to the bread. It contained no additives to aid in loaf release.

Pan greases have been made by adding beeswax, rice bran wax, paraffin, oxystearin, lecithin, or combinations of these to liquid vegetable oil. One pan grease contained aluminum stearate as a thickener. This was done to enable the grease to remain where it was sprayed without puddling or running off of the pan.

MISCELLANEOUS SPECIAL SHORTENINGS

Icing Shortenings

Icing shortenings are designed to cream with sugar and water and to incorporate large volumes of air. The efficiency of the shortening is measured not only by the lowest specific gravity obtainable but also by the length of time required to reach it. The icing must also be able to retain its low gravity without leakage of water. Performance tests are easily designed to evaluate these factors.

As with all shortenings used for creaming, beta-prime hardfat plasticizers are essential for icing shortening formulation. Softer hardfats are required for icings than for cake shortenings. Specifically, cottonseed oil hydrogenated to 50–52 titer hardness is best for icings, to 58–60 titer is best for cakes. Shortening plasticized with the softer hardfat has a narrower plastic range than one formulated with the harder material. The wider plastic range seems to benefit cake baking performance.

Hardness of the monoglycerides used for icing shortening preparation seems to be even more critical than hardness of the plasticizer. Monoglycerides made from unhydrogenated vegetable oils give the best performance in icings. Shortenings containing monoglycerides of fully hydrogenated fats will often produce icings with practically no aeration.

Polyglycerol esters of oleic acid or of mixed fatty acids from unhydrogenated cottonseed oil are particularly active in promoting aeration of icings. Polysorbates are also excellent in icing shortening emulsification.

An average icing shortening formulation consists of 17% cottonseed oil flakes (50–52 titer), 2% polyglycerol ester, or 3% soft monoglyceride (by analysis). The balance of the formulation is the base oil. This is either partially hydrogenated soybean oil with an iodine value of about 80 or interesterified lard (Howard and Konen 1965). Obviously, the exact formulation will vary with each producer. As new emulsifiers are developed, they will replace or augment those given above.

A fluid icing shortening has been developed which consists of 94% cottonseed oil, 4% hydrogenated lecithin, and 2% stearic acid (Howard and Konen 1965).

Cookie Fillers

Sandwich cookies contain a creme filler made by mixing powdered sugar and shortening into a stiff paste. Lecithin is usually added to aid in mixing. Lecithin promotes spreading of the fat over the sugar surfaces.

The shortening must be fairly hard in order to obtain rapid setting up of the filler after it is deposited on the cookie. The shortening contains no emulsifier. If lecithin is used, it is added as such to the filler formulation at time of preparation.

Lard hydrogenated to 50–55 iodine value or soybean oil hardened to 65–70 iodine value are commonly used creme filler fats. From 3 to 5% hardfat is usually added to the base to shorten setting time. The cookies are cooled before the filler is deposited.

In one case, where the cookies were not adequately cooled before deposition of the filler, residual heat caused the filler to melt after the cookie sandwiches were packaged. An all-purpose unemulsified shortening was substituted for the normal filler fat. The high level of hardfat in this wide plastic range shortening was able to tolerate the temperature rise without melting.

BIBLIOGRAPHY

BALDWIN, R. R. et al. 1965. Fat systems for continuous mix bread. Cereal Sci. Today 10, 452–457.

BEDENK, W. T. 1962. Shortening in dry prepared culinary mixes. U.S. Pat. 3,037,864. June 5.

HOWARD, N. B., and KONEN, P. M. 1965. Fluid shortening for cream icings. U.S. Pat. 3,208,857. Sept. 28.

PARKER, H. K. 1965. Continuous mix baking to date. Cereal Sci. Today 10, 272–276.

SCHWAIN, F. R. 1965. Fifty years of progress in fats and oils. Cereal Sci. Today 10, 277–283, 353.

CHAPTER 7

Frying Shortenings and Their Utilization

Frying is one of the most important methods of food preparation. Yet it is the method which is least understood by persons who earn their livelihood from foods. Frying is a deceptively simple operation. It is, therefore, subject to much abuse.

Restaurants practice both deep fat and pan frying. Industrial frying is a continuous deep fat process. Specialized shortenings are marketed for each type of frying. The shortening is not a cure-all, however. Good frying practice is essential.

INSTITUTIONAL FRYING

Fried foods have always been popular items on the restaurant menu. When well-prepared, the foods are crisp on the outside and moist and tender on the inside. They are flavorful with an appealing color and odor. Although the food is in contact with shortening during frying, it should not have a greasy appearance or taste. The surface of the food should appear dry with no excess oil adhering to it.

Deep Fat Frying

In deep fat frying the food is completely surrounded by the frying fat. The fat acts as a heat transfer medium. As such, it is more efficient than dry heat of the oven. It is more rapid than boiling in water as the temperatures used in frying are higher, causing more rapid heat penetration. The chef would prefer frying if for no other reason than the economies offered by short cooking times.

The shortening is more than a means of transferring heat to the food. The fat reacts with the protein and carbohydrate components of the food, developing unique flavors and odors which have definite appeal to the consumer.

Frying is carried out correctly when the food pieces start to evolve bubbles of moisture shortly after being immersed in the hot oil. Steam must boil off continuously throughout the frying operation until the food is removed from the fry kettle. The oil temperature must be above the boiling point of water but need not be very high to result in actual frying. The temperatures normally used range from 325° to 385°F, depending on the food to be fried. Although lower temperatures will bring about frying, the length of time required for thorough cooking would make the opera-

TABLE 19

DEEP FAT FRYING GUIDE

Food	Coating	Temp., °F	Time, Min
Meat			
Cutlets	Batter and breading	360	5–8
Chicken fried steak	Batter	350	5–8
Chops, very lean	Breading	350	5–8
Seafood			
Fish cakes	Breading	370	2–3
Clams	Batter	365	1–3
Fillets	Batter or breading	370	3–5
Oysters	Batter or breading	360	1–4
Scallops	Breading	360	3–5
Shrimp	Batter and breading	365	4–6
Smelts	Breading	370	4–6
Turnovers			
Turkey, chicken, tuna	Dumpling dough	380	5–7
Poultry			
Chicken, large pieces	Batter or breading	350	10–15
Chicken, small pieces	Batter and breading	365	7–10
Chicken, precooked	Batter and breading	365	3–5
Turkey, small pieces	Batter and breading	365	9–10
Vegetables			
Eggplant, sliced	Breading	365	5–7
Onions, sliced	Light batter	365	2–3
Asparagus, precooked	Batter or breading	365	2–3
Cauliflower florets, pre-cooked	Batter or breading	365	2–3
Potatoes, $1/2$-in. sq, one operation		370	6–9
Potatoes, $1/2$-in. sq, blanch		370	4–6
Potatoes, $1/2$-in. sq, brown		370	2–3
Potato chips		370	2–3
Potatoes, julienne, $1/4$-in. sq		370	3–6
Doughnuts			
Cake type		375	1.5–2
Yeast-raised		375	2–2.5
Dumplings			
Turkey, chicken	Dumpling dough	380	5–7
Fruit	Dumpling dough	380	10–12
Fritters			
Fruit	Batter	360	3–5
Vegetable		350	5–8
Miscellaneous			
Chinese noodles	Batter	375	1–2
Croquettes	Breading	375	2–3
French toast	Batter	375	2–3
French-fried rice, pre-cooked		375	2–3
Nuts			
Almonds		240	4–6
Cashews		275	3–5
Peanuts		350	3–5
Blanched peanuts		300	3–5

Source: Gidden-Durkee Div., SCM Corp.

tion unsuitable for commerical restaurant use. Temperatures above 385°F would result in rapid degradation of the frying fat. In addition, excessively high temperatures would result in undercooked food on the interior of the piece while the exterior might well be burned. Table 19 lists recommended times and temperatures for the most efficient frying of a number of foods.

Commercial fry kettles are normally equipped with thermostatic controls. Thermostats have a tendency to drift. It is, therefore, advisable to check the actual fat temperature with a thermometer at least weekly to correct for this drift.

As a general rule, small pieces of food are fried quickly at high temperatures. Large pieces which require a longer time for thorough penetration of heat are cooked at a lower temperature. This prevents overcooking or burning of the food surface. It seems obvious that food pieces that differ markedly in size should not be fried in the same kettle at the same time.

Continuous evolution of steam from the food throughout frying is essential as this indicates that the inside of the food is at a higher pressure than the oil in the kettle. This keeps shortening from penetrating the food surface. If the internal pressure were to drop during frying, as observed by cessation of boiling action, the shortening would seep into the food and cause it to be greasy.

The preparation of nongreasy fried foods does not depend on the shortening that is used. It is a function of the heat balance in the frying system. It is a mistake to attempt to correct a greasy food problem by changing shortenings. The problem will persist.

Heat balance in the fryer refers to the relationship between heat input and heat requirement. The ability of the fryer to heat the shortening is measured in Btu per hour. This is inherent in the fryer design and is the only real limitation on fryer capacity. The requirement for heat is controlled by the fry cook. Satisfactory performance of the fry kettle relies on keeping the demand for heat below that which the heating system can supply.

The heat requirement, or fryer load, is a function of how rapidly moisture is to be removed from the food. Oil temperature drops when cold food is placed in it. Evaporation of moisture by boiling cools the fat further. The system must be able to replace the heat at a slightly higher rate than at which it is lost. This is to maintain an increasing oil temperature during frying and thereby a continuous evolution of steam.

Fryer load is determined by the relationship between the amount of moisture to be removed and the availability of the moisture. A fry basket of large pieces of food may have as much total moisture as a load of small pieces. Surface area of the large piece is small in relation to that moisture content. Small pieces have a large surface area per unit weight. Heat penetration into the small piece is rapid. Moisture loss is also rapid. Small pieces would require the higher heat recovery rate.

Fryer capacity will vary with the food to be fried. A well-designed fryer will have baskets of such size that it will not be overloaded with any food if the individual basket is not filled over half way.

It is, of course, possible to overload a fryer even then. Foods to be fried should be well-drained of excess water. Fine cut foods, such as Julienne potatoes can retain a large volume of free water after being rinsed. Poor drainage will carry that water into the fry kettle and cause an overload. If the basket is also filled too far, the evolution of steam bubbles can be of sufficient volume to cause the oil to boil over the sides of the fryer.

Fryer overload is most common during rush hours. The temptation is to put too much food in the fry basket. Then the thermostat is set higher to compensate for the added load and to cook the food faster. In the extreme case this results in overcooking the surface of the food, leaving the inside underdone. If the thermostat is set lower to reduce burning, the fryer cannot recover heat loss fast enough. The food becomes soggy and greasy. The correct solution is to install additional fry kettles or to replace undersized kettles with larger models. Old fryers which hold a specific volume of shortening may not have a fast enough recovery rate. Newer, quick-recovery fry kettles have a higher frying capacity for the same volume of fat as they have larger heating elements or burners.

Quick-recovery fryers have another advantage. It is sometimes difficult to predict when a fryer will be needed to fill an order. A kettle with a rapid heating rate can be brought up to frying temperature quickly, making it unnecessary to hold the frying oil at too high a temperature during slow periods. A temperature range of 200°–250°F is recommended for fryers on standby. This retards oil breakdown.

Oils deteriorate more or less slowly when heated. Figure 14 outlines the types of deterioration which take place. Fats held at frying temperatures or higher but without frying in them break down rapidly and soon become unusable for frying. When foods are fried in a fat, the evaporating steam strips off some of the decomposition products and delays deterioration of the fat. Lowering the temperature of the shortening when it is not in use also delays deterioration of the fat. The adequacy of commercial frying methods in protecting fat quality has been demonstrated (Jacobson 1967).

Turnover rate is probably the most important factor in retaining frying fats in a satisfactory condition. Turnover rate refers to the proportion of shortening which is to be added to the kettle daily to maintain the fat level in the fryer at the "full" mark. Makeup fat should amount to 15–

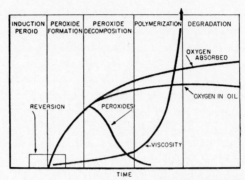

From Perkins (1967)

FIG. 14. STAGES OF PROGRESSIVE DETERIORATION
OF FATS THROUGH OXIDATION

Copyright © (1967) by Institute of Food
Technologists

25% of the total capacity of the kettle. If an insufficient amount of food is fried daily, resulting in too low a turnover rate, it is advisable to discard some of the fat in the kettle in order to increase the amount of fresh fat to be added.

Shortening loss from the fryer is due to absorption of fat by the food being fried. Some is also lost during filtration of the fryer fat which should be done at the end of each day's operation. Filtration is used to remove particles of food, usually crumbs of breading, slivers of potato, etc., which have accumulated during frying. Failure to remove these particles could result in their burning and depositing on the food as black specks. The burnt particles could also develop undesirable flavors and odors and increase the deterioration of the frying fat. A fry kettle with a cold zone on its bottom helps remove food particles from the hot fat during frying.

While badly deteriorated shortening is not a common occurrence in a good frying operation, it is not unknown. Broken down fat is dark, viscous, burnt in odor and flavor. It tends to smoke and foam badly. More important, foods fried in such fats are also dark and bad tasting.

Too low a turnover rate is only one possible cause of this type of fat breakdown. The presence of copper in the system is another major contributing factor. Fryer manufacturers are aware of the prooxidant activity of copper on fats. Unfortunately equipment repairs may be made by uninformed workmen. Brazing of broken baskets and use of copper, brass, or bronze pet cocks, filter screens, piping or heating ducts are typical sources of trouble. Stainless steel is the preferred metal for fryer construction.

Shortening which has broken down must be discarded and replaced with fresh fat. The color of food fried in unused fat is usually too light. Longer frying time does not result in darkening to a proper fried color. Shortenings contribute deeper coloration to foods as the fats become used in previous frying. Some shortenings impart appropriate color to foods fried in them sooner than do other fats. It would seem that oxidation of fats or other components of the shortening is involved in color development in foods. The more unsaturated fats and oils seem to develop the ability to color foods during frying faster than do the harder shortening compounds.

Free fatty acids form during frying due to presence of moisture in the foods. A free fatty acid level of 0.5–0.8% is normal for frying fats in use. If the frying fat is not hot enough to drive the water from the food at a high enough rate, the free fatty acid level will increase too rapidly. While this may not affect flavor of the fried food, the shortening will begin to smoke at too low a temperature. Anything which restricts loss of water from the fry kettle will have the same effect. Water dripping from a poorly vented hood into the frying fat is a good example. Fry kettles should never be covered during frying for the same reason. Covering the kettle during idle periods to restrict exposure of the fat to air, dust, or other foreign materials is a good idea.

Proper turnover rate will keep the frying fat at a sufficiently low free fatty acid level. It will also replace silicones which are lost slowly by adhering to the food. A high enough turnover rate, however, would make the use of silicones unnecessary. Foaming occurs when the fat has polymerized to a certain degree. Silicones break the foam as it forms. Good fat turnover would maintain a polymer level below that which would cause foam formation.

Foaming can result from reasons other than frying fat breakdown. The frying of food dipped in egg batter or doughnuts containing a high level of egg solids can cause early foaming from lecithin leaching into the shortening. Chicken, meats, and fish which are high in fat content exchange these fats with those of the frying medium. A high stability hydrogenated shortening can become contaminated with low stability fat and a number of other unstable compounds accompanying it.

Foam from fat breakdown or slow accumulation of foreign matter cannot occur in fresh shortening. Occasionally a shortening will foam immediately on being added to the fry kettle. If it is makeup fat being added to older shortening in the fryer, the new fat may have an excess amount of suspended soaps. If the fat foams on being added to a recently cleaned fry kettle the fault is probably in the cleaning operation.

The kettle and baskets should be cleaned daily to rid them of accumulated gums of fat polymers. This is done by boiling the equipment in a solution of alkaline cleaning compound. The residual alkali should be flushed out with clear water followed by dilute acid, usually vinegar. The vinegar solution is then rinsed away with more clear water. Vinegar removes soaps which form from the fats during boil-out with alkali. If the vinegar rinse is omitted, residual soaps can cause foaming when frying is resumed.

Cleaning compounds are specially formulated for fryer cleaning. Some cleaners formulated for other uses may contain nonionic wetting agents. These are difficult to rinse away with water and are not destroyed by vinegar. Residual wetting agents will cause severe foaming in frying. Care should be taken in selection of a proper cleaner.

A variety of shortenings are suitable for deep-fat frying purposes. These range from unhydrogenated vegetable oils and meat fats to specially hardened shortening compounds. Thompson *et al.* (1967) have shown that the frying life of these shortenings is not dependent on the amount of unsaturation but on how the fats are used. Fried foods are usually served hot, doughnuts being the major exception. The frying fat does not require unusual stability as foods fried in it are not stored before being eaten.

Shortening selection might well depend on turnover rate experienced by the individual restaurant operation. A low rate would require the use of hydrogenated fats. Hydrogenated vegetable oils with the highest stability are the narrow plastic range type with little or no added hardfats. Hydrogenated lard is also quite stable. All-purpose unemulsified shortenings are next in decreasing order of stability. The special frying shortenings usually have a higher silicone content than the all-purpose fats. In fact, some shortening suppliers do not put silicones in their all-purpose brands.

A high shortening turnover makes the use of unhydrogenated vegetable oils and deodorized and stabilized lard feasible as frying fats. Cottonseed, peanut, or corn oils are preferred to unhydrogenated soybean oil. Liquid soybean oil develops a fishy odor on frying which is highly objectionable. Soybean oil has been used in kitchens with exceptional ventilation to remove frying odors. Food fried in this oil does not seem to have an off-flavor when served shortly after its preparation.

Lightly hydrogenated and winterized soybean oils are available for frying purposes (Haynes 1966). Such oils are somewhat similar in stability to unhydrogenated cottonseed oil. The different vegetable oils have their characteristic odors on being heated to frying temperatures. This seems to affect only the kitchen area. Frying odors should not reach the serving area no matter how bland they may seem. Room odors are not always obvious to persons in the room during frying. The odors become evident when a person leaves the room and returns shortly thereafter. The odor may also seem stronger to people in the next room.

Undeodorized lard has been used for deep fat frying. It should be clarified to remove protein residues as these would burn on frying and cause development of off-flavors and -odors which could be carried over into the food. Undeodorized lard may also have a high free fatty acid content which would cause smoking at a low temperature, e.g., 325°F.

A recent frying shortening development is the pourable type with relatively high stability. It is prepared by fractionation of hydrogenated soybean oil with an iodine value of 65–70 (Cochran *et al.* 1961; Simmons *et al.* 1968). The resulting oil is liquid at room temperature but is almost as stable against oxidation as the oil stock from which it is made. The advantages of liquid frying fats are that they can be poured into the fry kettle when cold, heated without the danger of burning on hot coils, cooled without solidifying and filtered while cold. This last advantage is of prime importance as a safety feature. Filtering of hot shortening has always been a hazardous operation.

Some pourable frying shortenings are in fluid form in that they contain a small amount of suspended solid fats. They have one possible advantage over clear liquid shortening in that they would tend to retain added silicones in suspension with the fat solids. Clear liquid oils seem to lose silicones while in storage, probably by migration of the silicone to the container wall.

Fluid shortening, however, does contain solid fats which settle out on cooling. This would remove silicones as well as fat crystals from the main body of fat in the kettle. If the kettle has a cold zone, it is necessary to stir up the fat which settles in the cold zone on reheating the kettle in order to resuspend the silicones. Fluid fats must be filtered at a slightly higher temperature than liquid oils, but not at a temperature which would be considered as a safety hazard.

Certain types of shortening are completely unsuitable for deep fat frying. Emulsified shortenings would smoke at too low a temperature and might cause foaming. Pan and grill fats contain lecithin. This material darkens excessively on being heated to frying temperature. It would also cause a high degree of foaming. Shortenings containing mixtures of coconut and nonlauric acid oils would foam badly. Colors and flavors break down quickly at frying temperature and are wasted in the deep-fat fryer.

Pan and Grill Frying

Shortenings in pan and grill frying have two functions. One is to contribute or aid in developing of flavor and color in food. The other is to prevent the food from sticking to the hot cooking surface. Most of the color and flavor comes from the reaction between protein, carbohydrate, and fat and their breakdown products at frying temperature.

Pan frying is more of an art than is deep fat frying. In the latter, time and temperature are measurable and subject to some mechanical control. In pan frying, the time and temperature are usually judged by the chef. If it is too low, the food will cook too slowly. Although in many cases this is preferable from a palatability standpoint, the chef cannot afford the time required. The temperature is therefore increased.

If the grill is too hot, the food surface will burn while the interior will be barely cooked and may even be slightly underdone.

In pan and grill frying the fats are exposed to high heat in thin films which are also exposed to oxygen in the atmosphere. This tends to harden the oils into a varnish-like layer on the grill surface. This layer should be scraped or scoured off regularly to prevent buildup on the frying surfaces. The softer resinous deposits can contribute paint-like flavors and odors to the fried foods.

Any unemulsified shortening can be used for pan frying. Specialized shortenings have been designed to give improved performance over the ordinary fats and oils. Lecithin is a common component of pan and grill shortenings. It is one of the best antistick compounds. Lecithin also causes foaming which prevents spattering of water droplets during frying. Coconut oil is also added to some shortenings to promote foaming. Obviously, silicone antifoam agents are not desirable in pan and grill frying. Some multipurpose frying fats for small restaurants may contain silicones for deep-fat frying but rely on the fat component itself to provide antisticking properties for pan and grill use.

Buttery colors and flavors are often added to pan and grill shortenings. The colors and flavors are of the heat stable variety. They are eventually destroyed by standing on the hot grill surface. If cold food is placed on the grill soon enough after application of the shortening, sufficient color and flavor adheres to the food to be of value. Colored and flavored shortenings are sometimes brushed on toast in place of butter or margarine.

Most pan and grill shortenings are pourable for convenience. A few are soft solids which can be brushed on the grill surface. A process for production of one such semisolid fat has been described by Reid (1969).

INDUSTRIAL FRYING

Equipment

A number of foods are fried on a large enough scale to be regarded as industrial. Frying of this dimension is usually done on a continuous basis in fryers especially designed for the purpose. Figure 15 shows such a fryer. Food items are prepared for frying by mechanical means as far as is possible with the particular food concerned. Hand labor is used primarily to sort out culls. Operators set up the equipment and make sure that it continues to operate properly through the day.

In general, food is fed to the fryer by a continuous belt which ends at the edge of the fry kettle. The pieces of food tumble into the hot oil and are moved through the fryer by chain belt or by laterally placed bars moving the length of the fryer. The other end of the fry kettle is equipped

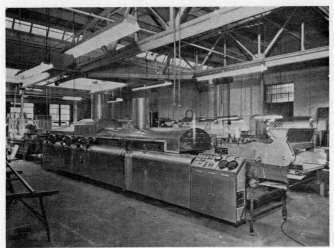

Courtesy of J. D. Ferry Co., Inc.

FIG. 15. INDUSTRIAL POTATO CHIP FRYER

with a chain belt ramp which removes the fried food for further processing.

Fryer design varies according to the food to be fried. Some equipment is designed to keep the food submerged if it tends to float or to turn the pieces over at some point so that both sides are cooked about equally. Fryers may have two separate kettles connected by a chain belt ramp for transfer of food. Frying oil in each kettle is maintained at a different temperature. The second kettle is usually the hotter one to obtain constant loss of moisture from the food throughout the frying operation. The moving ramp is designed to turn the food pieces over as they drop from the belt into the next kettle. Doughnut fryers have a paddle-like flipper timed to turn the doughnuts over at the proper degree of doneness on one side.

Kettles are designed to be either direct-fired or heated through a side-arm heat exchanger (Rock and Roth 1964). In the external heater system, oil is pumped from the kettle through the heater and back to the kettle. The oil is removed from the kettle at the end from which the finished food is removed. The reheated oil is returned to the kettle at the point where the raw food enters, and in some cases, through additional entrance ports part way down the fry kettle. This is especially true where the fryer has two separate kettles.

Direct-fired kettles can develop areas of burned-on fats where the heat is too intense and the flow of oil is too slow. Sidearm heaters can cause burned fats if they have parallel heating coils through which the oil flows. Blockage of one heating coil can go undetected until it burns through or becomes unblocked, permitting the burned oil to return to the fry kettle.

External heating systems can be the source of another problem, that caused by aeration. Fats should never be allowed to gush through air or have air whipped into them. This is especially true where the fats are near or at frying temperatures. Aeration can occur through leaks in piping on the suction side of pumps. The pump packing gland may leak and draw air into the line. Oil is often permitted to drop from the end of a pipe into a catch basin or into the fry kettle. The oil is usually under pressure from a pump and may be returning directly from the heat exchanger. Aeration caused by this gushing of hot oil can bring about its rapid breakdown.

The oil in continuous frying systems must be filtered continuously. Enclosed pressure filters are most common. The hermetically sealed centrifuge is also used to clarify the oil. A metal screen is placed ahead of the centrifuge to remove large pieces of solid matter. The fines are then removed by the centrifuge. Occasionally the centrifuge discharge has been allowed to gush into an open catch basin. The outlet pipe in such cases should be extended to reach below the oil level in a covered basin. The oil depth should be sufficient to prevent aeration of the oil by turbulence. Open filters, especially with flow patterns which permit cascading of hot oil, should be avoided.

Oil temperature and frying time are determined by the food to be cooked. The temperature is controlled by thermostats. Timing is adjusted by varying the speed of the belt or push-bars moving through the kettle. The rate of food movement through the entire processing system is governed by the slowest operation. If preparation of the food prior to frying becomes slower than normal for any reason, movement through the fryer should also be slowed down. Oil temperature should be reduced accordingly. The guideline is that the food must evolve steam or boil continuously through the entire length of the kettle and be sizzling on removal from the hot fat. The food must also be cooked to the desired degree of doneness. This does not mean that the food must be completely cooked. Oil blanched French-fried potatoes, for example, are only half cooked at the end of the frying operation.

Continual frying is essential in continuous operations. The blanket of steam evolved from the frying food protects the hot oil from excessive oxidation. If a slowdown occurs in feed rate to the fryer, it is better to slow the fryer throughput rate and maintain a steam blanket than to hold the food and fry on an intermittent basis.

Continuous fry kettles are covered with a vented hood. This not only keeps a blanket of steam over the oil, it also keeps steam out of the room. The hood should be raised from the oil surface to allow for escape of steam through the flue. Home-made fry kettles have occasionally had insufficient space between oil surface and hood. The steam would condense and run back into the oil causing excess free fatty acid development in the oil.

The hood should have a gutter around it. The flue should be provided with a baffle. This is to prevent steam condensate from getting into the oil. Poor venting, especially noticeable on days of low atmospheric pressure, causes steam to fill the area around the fry kettle.

In such cases, steam condensate has been observed dripping back from overhead pipes, light fixtures, etc., onto the food and into the oil in the kettle. Forced draft ventilation may be required in such an event.

While continuous frying is a deep-fat operation, the depth of oil should be no greater than is necessary to cover the food adequately. This keeps the fat turnover rate at the highest possible level. The total amount of fat removal is proportional to the amount of food passing through the fryer. The percentage removed is increased as the amount of fat in the kettle is decreased. A well designed operation results in complete turnover of the shortening in a kettle in about 16 hr. This is quite different from restaurant frying which hopefully attains complete turnover in 4 to 7 days.

Large-scale industrial frying requires the same care in cleaning of equipment as was outlined for small scale restaurant frying. The same is true for the harmful effects of copper contamination on the oil.

Potato Chips

Potato chips, potato puffs (extruded potato flour paste), corn chips (extruded corn meal paste), and similar products are fried and packaged to be consumed within a few days after preparation. The frying shortening need not have a high stability. Cottonseed cooking oil is used frequently where a bright finish is desired on the chip. Corn or peanut oils are also satisfactory but are usually more costly. Unhydrogenated soy-bean and safflower oils are too unstable and revert to give the chip an objectionable flavor. Slightly hardened soybean oil (iodine value 105–110) is sometimes used, either with or without winterization. It also re-verts in flavor but is not as objectionable as unhydrogenated soybean oil. A blend of 15–25% cottonseed oil and 75–85% slightly hardened soybean oil is more acceptable since the cottonseed oil contributes sufficient flavor to mask that of the soybean oil.

Chips with a dull finish are obtained by frying in partially hydrogenated oil with an iodine value range of 70–75. There is no detectable flavor difference between soybean and other oils at this level of hardness. Soy-bean oil is preferred as having the lowest cost.

Flavor stability of the chip is enhanced by the use of BHA and BHT in the frying oil. Propyl gallate does not carry through to the chip but is lost in frying. Antioxidants are occasionally sprinkled on the finished chip by mixing them with the salt. In such a case propyl gallate would have antioxidant activity along with BHA and BHT. Antioxidants are not added to the shortening if they are to be added later with the salt.

The amount of oil absorbed by the chip, usually 32–45% of the chip weight depends on the frying method and the potato or other material used. It does not depend on the type of shortening in which the chip is fried.

Frying chips to complete dryness will cause excessive darkening where the potato has a high reducing sugar content. This is prevented by fry-ing to 6–10% moisture level and finishing the chip to 0.5–0.75% moisture content outside the fry kettle. A hot air tunnel at 250°F, infrared lamps, or a microwave oven can be used for this operation.

Canned Fried Potatoes

Canned Julienne or shoestring potatoes are fried in much the same way as potato chips, but the frying oil is usually cottonseed oil hardened to an iodine value of 70–75. Some producers prefer hydrogenated peanut oil. They feel that these oils develop less off-flavors than does soybean oil of equal hardness. The canned product requires a long shelf-life in con-trast to bagged chips.

The odor of air trapped in the can may differ with the various oils but it is questionable whether any detectable flavor differences exist in the potato product. Potato flavor is fairly definite and would tend to mask oil flavors of low intensity. If frying oils of higher iodine values were to be used, off-odors of sufficient strength might develop to make the selec-tion of oil source more critical than it is.

Snacks

A number of snack items have appeared on the market in the past few years. Many of these are baked, but some are fried. The usual frying medium is hydrogenated coconut oil. This oil melts below body temperature but still has an iodine value of about one unit. It is very stable against oxidative rancidity.

Stability is a necessity in these snack foods as they are distributed nationally in pasteboard boxes. This could not be done with potato chips.

Antioxidants are added to the snack foods to protect them from oxidation. The frying oil does not require protection. Natural oils in the other ingredients, especially traces of corn gluten oil in corn flour, apparently cause oxidative rancidity and must be stabilized.

Nuts

Nut meats are often fried, salted, and canned. They are normally fried in unhydrogenated coconut oil. This oil has a low viscosity due to its high content of short chain fatty acids. This makes the nuts seem less greasy. Coconut oil has a low iodine value and is fairly stable. However, the natural oil in nut meats is high in polyunsaturated fatty acids. Fried nuts must, therefore, be packed in vacuum to limit the oxygen content in the can. Antioxidants are often used to scavenge the last traces of oxygen from the can. Small packages of fried nut meats are often sold without such protection if rapid market turnover can be anticipated.

Doughnuts

As with some commercial cakes, doughnuts are also often packaged and distributed over distances of several hundred miles from the bakeshop. They must have a shelf-life of a few days before becoming stale. The shortening in which doughnuts are fried need not be highly stable but must resist rancidity for at least as long a time as the doughnut remains fresh.

Large-scale doughnut fryers are timed automatically, not only to fry but to cool the doughnut. At some point during the cooling period, the doughnuts may be dusted with powdered sugar or glazed. The time actually required depends on the overall fryer design. The shortening must be firm enough at this time to pick up and hold the covering properly. If the shortening is too oily because it has not set up sufficiently, the glaze will not stick and powdered sugar will become grease-soaked. If the shortening has set up completely, the glaze will probable adhere satisfactorily, but the powdered sugar will not.

The length of time required for a shortening to harden to the desired degree is more a function of crystal structure than of amount of hard fat in the shortening. Beta-prime hardfats set up more quickly than do beta crystal fats. An attempt to slow the setting time of a beta-prime shortening by reducing the hardfat content could result in having too little hardfat. Consequently the shortening will be too soft and will cause grease soakage no matter how long the doughnut has been cooled.

Shortenings most commonly used for doughnut frying are the unemulsified all-purpose shortenings and deodorized, stabilized lard containing 8–12% fully hydrogenated lard flakes.

Frozen Breaded Foods

Breaded meat patties, fish, and seafood are often fried, packaged, and then frozen for storage and distribution. They are usually completely cooked so that they need only be reheated to be consumed. Reheating may be by deep fat or pan frying or by dry heat in the oven.

The main problem seems to be in selection of an appropriate breading. Different breading components darken at various rates, probably depending on their sugar content. Corn flour hardly darkens at all even after lengthy exposure to frying heat. Crushed corn flakes contain added sugar and darken quickly at frying temperatures. Small or thin food pieces should have a breading which darkens quickly as these foods cook in a short time. Large pieces require a breading which darkens slowly so as not to become too dark by the time the food is cooked.

On being reheated in the oven, some of the fatty foods, such as meat patties, may tend to leak oil and appear quite greasy. The dipping of food in a 1–2% aqueous methylcellulose solution before breading has been recommended as a means of reducing fat absorption from the fryer and subsequent leakage from the food during reheating (Scheffel and Klis 1965).

Stability of the frying fat is of no great concern in frying foods to be frozen. The deterioration of fats at freezer temperatures is not significant in amount to be detectable by chemical or organoleptic means. Unhydrogenated oils are not used unless they contain some added harder fat components. This is to eliminate an oily appearance of food in the frozen state.

All-purpose shortenings are not ordinarily satisfactory as they contain a high level of hardfat which can appear as blobs of waxy material in the frozen package. Narrow plastic range frying fats or unhydrogenated, deodorized lard with perhaps 5–8% lard flakes seem to be the preferred shortenings for frying breaded foods.

Oil-blanched French-fried Potatoes

French-fried potatoes are partially cooked or blanched in oil for distribution to the customer who then finishes the frying operation. The restaurant chef saves time in that the potatoes are not only peeled and cut but take less time to fry than a completely raw potato. The housewife has the added convenience of being able to finish cooking in an oven so that she may serve French-fries without using a deep fat fryer.

The oil-blanching operation in itself is similar to that used for potato chips. The temperatures and times used are lower, however, to obtain only partial doneness of the potato.

The potatoes are prepared for frying by lye-peeling, washing, cutting, and preblanching in hot water or steam. The potato pieces may then be dipped in a dextrose solution or a yellow water-soluble dye solution to help develop a finished potato with a desirable color. Some potato varieties are sufficiently self-coloring not to require this step. Excess water should be shaken and even blown off of the potato mass in order to avoid overloading the fry kettle.

After the potato mass is removed from the fryer it should again be shaken and air-blown to remove as much of the adhering frying oils as possible. The potatoes are then moved slowly through a cold air chamber before entering a blast freezer. The frozen potatoes are finally packaged and shipped.

Some oil blanchers merely refrigerate their potatoes and do not market a frozen product. Refrigerated blanched potatoes must be used within about one week as they will spoil after this length of time.

As with frozen breaded products, shortening stability for potato frying is not an important factor. It has been found, however, that unhydrogenated oils tend to make the potato look and feel greasy to the fry cook, especially since he often removes potatoes from the package by hand. Hard shortenings with melting points over 95°F set up too quickly in the precooler and freezer causing the potato pieces to cluster together. These clusters are difficult if not impossible to break up without also breaking the individual potato pieces. The preferred melting range for fats used in potato blanching is 80°–85°F.

The shortenings are usually hydrogenated soybean oil although hydrogenated cottonseed oil was once exclusively favored. The increasing disparity between soybean and cottonseed oil prices forced the potato blancher to make a complete reversal in his selection of oils.

Miscellaneous

Pressure Frying.—Frying of some foods, especially chicken, has taken on semiindustrial aspects. Although much frying of chicken is done in small take-out restaurants, the popularity of such food service is increasing to the extent that large batch fryers are presently being designed. The chicken is fried under pressure to shorten cooking time (Nelson 1965). The water content of pressure-fried chicken is usually higher than that fried in the open as only a sufficient amount of water is released as steam to obtain the necessary pressure.

The shortenings used are 65–70 iodine value hydrogenated soybean oil. Unfortunately, sufficient chicken fat leaches into the frying fat and increases its polyunsaturated fatty acid content. This shortens the frying life of the fat. A high turnover rate can offset its potential degradation.

Popcorn Oil.—Popcorn does not seem to fit any category. It is a snack item but is handled differently than the others. Corn is popped in a small layer of oil which adheres to the popped kernels. The oil is normally 76°F coconut oil containing a high level of heat stable carotenoid pigments.

BIBLIOGRAPHY

CHANG, S. S. 1967. Chemistry and technology of deep fat frying. An introduction. Food Technol. 21, 33–34.

COCHRAN, W. M., OTT, M. L., WONSIEWICZ, B. R., and ZWOLANEK, T. J. 1961. Domestic oil hard butters, coatings thereof and process for preparing said butters. U.S. Pat. 2,972,541. Feb. 21.

FEUSTEL, I. C., and KUENEMAN, R. W. 1967. Frozen French fries and other frozen potato products. In Potato Processing, 2nd Edition, W. F. Talburt, and O. Smith (Editors). Avi Publishing Co., Westport, Conn.

HAYNES, D. J. 1966. Develops flavor stable soybean oil. Food Process. 27, No. 5, 122–123.

JACOBSON, G. A. 1967. Quality control of commercial deep fat frying. Food Technol. 21, 147–152.

KAUNITZ, H. 1967. Nutritional aspects of thermally oxidized fats and oils. Food Technol. 21, 278–282.

NELSON, E. J. 1965. Method of deep fat frying and cooking. U.S. Pat. 3,194,662. July 13.

PERKINS, E. G. 1967. Formation of non-volatile decomposition products in heated fats and oils. Food Technol. 21, 611–616.

REID, E. J. 1969. Pan and grill fry shortening. U.S. Pat. 3,443,966. May 13.

ROBERTSON, C. J. 1967. The practice of deep fat frying. Food Technol. 21, 34–36.

ROCK, S. P., and ROTH, H. 1964. Factors affecting the rate of deterioration in the frying quality of fats. II. Type of heater and method of heating. J. Am. Oil Chemists' Soc. 41, 531–533.

SCHEFFEL, K. G., and KLIS, J. B. 1965. Use of methocel produces uniformly golden brown, less greasy fried foods. Food Process.-Marketing 26, No. 10, 104–106, 110, 112.

SIMMONS, R. O., REID, E. J., BLANKENSHIP, A. E., and MORGAN, P. W., Jr. 1968. Production of liquid shortening. U.S. Pat. 3,394,014. July 23.

SMITH, O. 1967. Potato chips. *In* Potato Processing, 2nd Edition, W. F. Talburt, and O. Smith (Editors). Avi Publishing Co., Westport, Conn.

THOMPSON, J. A. *et al.* 1967. A limited survey of fats and oils commercially used for deep fat frying. Food Technol. *21*, 405–407.

CHAPTER 8

Household Shortenings

Shortenings for household use are not formulated in the same way as those designed for commercial operations. While they serve similar functions, household shortenings have their own unique requirements.

SOLID SHORTENINGS

The old standard shortening, lard, is still available in retail markets. Some changes have been made in the product. It is now deodorized and stabilized. It is plasticized and filled into 1- and 3-lb paper cartons. Most lard was used for making pie crusts and for frying. The sale of lard in retail markets has dwindled with the advent of ready-to-bake frozen pies and biscuits. Probably most of the start-from-scratch baking is done today in small towns and rural areas.

The best known brands of solid shortening are designed for multipurpose uses. There is no accurate data available on what the housewife does with solid shortenings. At one time she used the product for baking cakes, cookies, and pastry, and for frying various foods. There is good reason to believe that the dry cake mix and frozen pie have all but eliminated the use of household shortenings for baking. The shortening producer, however, still formulates his product for cake baking as well as for frying. There is still a chance that many housewives will bake a cake from the shortening from time to time.

The shortening is formulated with a moderate level of monoglyceride for cake baking. The mono-diglyceride (40% mono content) is usually added at a level of 2–3%. This results in a monoglyceride level of 1.3–1.7% by analysis which includes the naturally occurring monoglycerides in the base oil. Commercial all-purpose shortenings contain about double this amount. The household cake, therefore, does not have the tenderness, moistness or antistaling quality of a well-made commercial cake.

The monoglyceride content of the household shortening is maintained at a low level so that it will not smoke excessively in frying. It would be impossible to fry in a commercial emulsified shortening. Polysorbate 60 has been evaluated for use in household shortenings (Purves *et al.* 1967). While cakes made from this shortening were more than satisfactory, polysorbate contributed a bitter flavor to fried foods and caused excessive foaming in the fry pan. The foam formation was desired as a means of reducing spattering but it was considered to be unsightly.

The presence of emulsifiers in household shortening has no significant effect on cookies or pastry baked with such shortening. It might affect the performance of cream icings, known as frosting to the homemaker. Most cream icings prepared in the home, however, are made with margarine or butter.

The same beta-prime hardfats are used in household as in commercial shortenings. Some all-vegetable brands use the same base oil, hydrogenated soybean oil with an iodine value of 80–85. The promotion of polyunsaturated fatty acid oils for nutritional purposes has led to the development of solid shortenings with a higher linoleic acid content than used previously (Anon. 1961). The harder shortening contains 8–18% linoleic acid. The softer product has 27–31% linoleic acid. This latter product is formulated from a soybean oil base hydrogenated to 105–110 iodine value. Cake baking performance may have suffered from this formula change.

Meat fat shortenings are usually made with interesterified lard as the base fat. Some may contain tallow and hydrogenated soybean oil as well. They are deodorized and stabilized with various blends of BHA, BHT, and propyl gallate. The BHA level is sometimes limited to 0.05% as it gives off a strong phenolic odor on frying which would be objectionable in a poorly ventilated kitchen.

Yellow colored shortenings were popular for a while. They contributed a pleasant yellow color to cakes, pastries, and pan fried foods. However, the color would disappear in the deep fat fryer. The fad for colored shortening was short-lived. Dead white shortening seems to be preferred.

Unemulsified household shortenings are available in limited areas, primarily in southern markets where frying is a popular form of cooking.

The tear-strip can with reclosable top was once standard. The shortening surface was smooth and shiny with a curlicue on the top, often touching the lid. The shortening was filled into the can and carried slowly on a delay table to the closing machine. The delay time was sufficient to insure that the shortening had solidified before it was spun on the can sealer. Adherence of shortening to the can lid was considered to be a serious defect. The tear-strip can was an expensive luxury.

The current container is an ordinary tin can with a plastic slip-on cover provided for reclosure. The can is completely filled with no free surface. This has eliminated need for the delay table.

Solid shortenings are packaged in 1-, 3-, 5- or 6-lb cans. They are occasionally packed in 1- and 3-lb paper cartons. The paper cartons are more commonly used for unemulsified shortenings.

The moisture content of canned shortening should not be more than 0.05%. Moisture above this level will corrode the interior of the can.

Levels in the area of 0.1% moisture have caused sufficient corrosion to bring about return of the product by the customer.

Whipped shortening is also being marketed. This product is fluffed with nitrogen to the extent that the normal 3-lb can size holds only 2 lb, 10 oz and a 1-lb can, only 14 oz (Clarke 1959; Dalziel and Dow 1959).

FLUID SHORTENINGS

Attempts have been made to market a household version of a lightly emulsified fluid shortening for general purpose use. The main feature was ease in measurement of the quantity of shortening and ease in creaming with sugar. The market tests were not successful (Anon. 1955).

LIQUID SHORTENINGS

Liquid household shortenings are essentially cooking or salad oils. They are sold under brand names. Winterized cottonseed and hydrogenated, winterized soybean oils are the most popular, followed by corn, peanut, and safflower oils. Blends of cottonseed with safflower or hydrogenated, winterized soybean oil are being marketed in some sections of the country when the crude oil price situation warrants it. Sesame seed, sunflower, rice bran, and citrus seed oils are sold on a local basis in areas where they are produced. Olive oil is sold to the gourmet market but is most frequently a deodorized product. Most grocery outlets do not stock virgin olive oil.

Most oils sold in retail markets are packaged under a nitrogen blanket to reduce the potential development of oxidative rancidity. Safflower and soybean oils are also protected with antioxidants, usually mixtures of BHA, BHT, and propyl gallate. The other oils have less tendency to develop objectionable off-flavors on reversion or oxidation.

Consumer preference has drifted toward blander and lighter colored oils over the years. At one time, corn oil was deliberately underbleached and underdeodorized to maintain a characteristic golden color and typical flavor. This practice has been discontinued.

Cottonseed oil is difficult to bleach at best. When the final color is judged to be too dark for the premium market, the oil is sold as a second grade product.

Soybean, cottonseed, and especially safflower oils sometimes have an excessive amount of chlorophyll which gives them a greenish cast. The oil producer usually tries to reduce the green color to a minimum level. Olive oil, however, is expected to be noticeably green.

An oil containing butter-like flavor and yellow color is currently being marketed for frying and seasoning purposes. It seems to be well-received but may prove to be another fad as was colored solid shortening.

Liquid oils are normally packaged in glass in 16-, 24-, 32-, and 48-oz sizes and in 1-gal. cans. Some odd sizes exist. Olive oil is often sold in 2-, 4-, 8-, and 16-oz bottles and 12- and 16-oz cans because of its high price. One gallon cans are also common. Blends of olive and cottonseed or soybean oils are offered in 1-gal. cans as a low cost substitute for 100% olive oil.

Fats and oils are sensitive to light which catalyzes oxidation reactions. It would be best to package liquid oil in dark glass or cans in order to protect the oil and extend its shelf-life. However, the consumer seems to prefer clear glass. Some oil brands are packaged in dark glass. Experience has shown that the sales of particular brands of oil have increased remarkably on changing their containers from colored to clear glass.

Polyethylene containers have proven to be unsatisfactory for packaging oil for the retail market. Polyethylene is permeable to oxygen. The turnover rate for oil sold for household use is sufficiently low that oil in polyethylene bottles usually becomes rancid before it can be consumed. Saran-coated polyethylene, polyvinyl chloride, and other oxygen impermeable plastic bottles seem to be as satisfactory as glass for packaging oil products.

PAN SPRAY

Lecithin is marketed in an aerosol spray dispenser as a means of obtaining oilless, low calorie frying. The film of lecithin on the pan surface does prevent sticking of food to the pan. The food, however, lacks the typical fried flavor formed by the interaction of hot oil with the food components. The effect seems to be more like waterless poaching than true frying.

On the other hand, the lecithin spray does make an excellent precoat for the pan to prevent sticking on frying with oils or margarine. A well-seasoned or tempered pan would not require such precoating if the pan surface were properly maintained. However, this is not always simple to do.

BIBLIOGRAPHY

ANON. 1955. New liquid shortening. Food Eng. 27, No. 8, 146.

ANON. 1961. Twice the polyunsaturates. Food Process. 22, No. 10, 9.

CLARKE, D. H. 1959. Manufacture of aerated shortening. U.S. Pat. 2,882,166. Apr. 14.

DALZIEL, W., and DOW, W. T. 1959. Process and apparatus for producing shortening. U.S. Pat. 2,882,165. Apr. 14.

PURVES, E. R., GOING, L. H., and DOBSON, R. D. 1967. Antispattering plastic shortening. U.S. Pat. 3,355,302. Nov. 28.

CHAPTER 9

Margarine

STANDARDS OF IDENTITY

Margarine is manufactured under U.S. FDA Standards of Identity. It is defined as a liquid or plastic food consisting of a mixture of fat and water. The minimum amount of fat permitted is 80% by weight. Any edible oil or fat may be used.

Many optional ingredients are permitted. Milk or ground soybeans were required prior to 1966 but were put on the optional list at that time. Margarines which contain milk solids form a curd on being melted. The water based product was developed to give a curd free margarine for frying purposes. Milk products, if used, may be fluid or reconstituted from condensed or dry form. Reconstituted dry milk is preferred as it is easier to handle and to store. Fluid milk must be delivered daily and be tested for proper pasteurization and lack of bacterial contamination before it can be used.

The milk product may be skim milk, sweet cream buttermilk, whole milk, or cream. It may be cultured with "harmless bacterial starter," one of several lactic acid producing bacteria, or be used without such starter. At one time, the starter was a source of naturally developed diacetyl and other butter-type flavor compounds. At present, any chemical approved for use as a food additive may be added to contribute a butter-like flavor to margarine. Flavors are usually purchased as branded items from flavor supply houses.

Any yellow pigment approved for food use by the U.S. FDA Food Additives Amendment may be added. The oil soluble carotenoids, beta carotene and bixin (annatto), are the two commonly used. Curcumin is often added to annatto to reduce the red character of the bixin pigment (Todd 1964). Colors are often purchased in batch-sized cans. Vitamin A is an optional ingredient. If it is to be added, it is usually blended with the color additive. Vitamin A must be incorporated at a level to give 15,000 IU per pound of margarine. Beta carotene has vitamin A activity so that the level of vitamin A alcohol or fatty acid ester is adjusted accordingly. Vitamin A is colorless. Vitamin D, temporarily withdrawn in 1965, was again permitted in margarine since 1968.

Mono- and diglycerides, lecithin, and sodium sulfoacetate derivatives of monoglycerides are permitted as emulsifiers. The total amount of these products must not exceed 0.5%. The function of the monoglyceride is to prevent leakage of water from the emulsion. The hardness of the mono-glyceride seems to be immaterial. The other emulsifiers on the optional list are added as antispattering agents for frying.

Citric acid, stearyl citrate, or isopropyl citrate may be added to the fat component as a metal scavenger. Calcium disodium EDTA is allowed in the aqueous phase for the same purpose, to prevent copper and iron con-taminants from acting as prooxidants for the fats and oils. These metals come both from equipment and from the various margarine components. Salt, an optional ingredient, is a major source of heavy metal contamina-tion. High purity salt is available, especially one pretreated with EDTA.

Sodium benzoate or benzoic acid and potassium sorbate are permitted at 0.1% as inhibitors for the growth of microorganisms. These com-pounds, as well as EDTA salts, are classified as "preservatives," a general term which has more legal than technical significance. Additional discus-sion of these materials will be found in Chap. 4 under preservatives.

Butter is permitted as an optional ingredient. It must appear, when used, in the ingredient phrase on the package label but can not be referred to by name in any other way, either on the package or in promotional ma-terial.

PROCESSING METHODS AND EQUIPMENT

In some ways, margarine can be considered as a shortening in emulsion form. This is especially so with certain bakery margarines which are pro-duced in the same manner as shortenings. It is a water-in-oil emulsion, taking its character from the continuous or oil phase.

Margarine production calls for intimately mixing the oil and aqueous phases and chilling the resulting emulsion.

Margarine may be manufactured in batch or continuous systems. The batch method produces a more uniform composition when proper care is taken. The Standards require a minimum of 80% fat. Analytical methods have an error of $\pm 0.2\%$ so that margarine should contain $80.2 \pm 0.2\%$ fat to meet the standards.

Direct injection proportioning systems are not sufficiently sensitive to meet these limits of error. For this reason proportioning pumps feed the margarine oil premixed with all the oil soluble components and the aque-ous phase containing all the water soluble components into a large agitated tank on a continuous basis. Variations in the proportioned feed are aver-aged out in this mixing tank. Another pump removes the emulsion con-tinuously, pumping it to a chilling machine.

The mixing tank or vat is called a churn although it bears no resem-blence to a butter churn. It is jacketed for temperature control and usually contains two high speed counter-rotating propellers. The emulsion formed is not stable in melted form. It begins to break within seconds after agitation ceases. The Votator or other internal chilling machine has good mixing action and will produce a uniform margarine if the emulsion being fed to it has not already separated into two phases. The distance between the final mixing tank and the chilling machine should be as short as is practical.

A flow diagram for the chilling of margarine is given in Fig. 16. There are many modifications of this process. The static B-unit produces a firm,

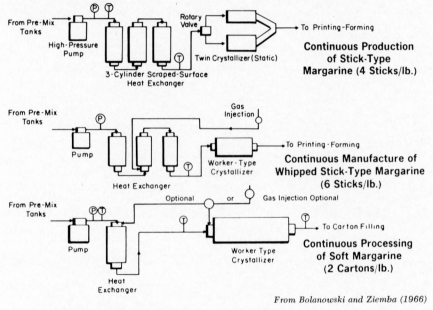

From Bolanowski and Ziemba (1966)

FIG. 16. SCHEMATIC DIAGRAMS OF MARGARINE CHILLING SYSTEMS

Reprinted from *Food Engineering*

nonaerated margarine. The chilled emulsion is extruded from it in the form of noodles. The noodles are then compressed into print form by the action of twin screw impellers feeding a print mold. The prints are most frequently 4 oz although $^1/_2$- and 1-lb prints are also available.

Whipped margarine requires a different type of B-unit (Fig. 16). Prints are extruded directly from it. Noodle plates and screw impellers would work the aerated emulsion excessively and squeeze gas from it. The amount of gas removed would be variable, making it impossible to control gas content. Since the print mold is adjusted by volume, a variable specific gravity would result in variable and possibly illegal print weight. Large diameter noodles and a piston fed print mold could probably be used in place of direct print extrusion to produce a satisfactory whipped margarine.

Some whipped margarine is filled into tubs. This requires a working B-unit as is used for shortening. Soft margarine is also filled this way (Fig. 16). Ordinary tub margarine may be slightly aerated if desired but not sufficiently so to be obviously whipped. Tub margarine is also produced without aeration.

Some pastry margarines are produced by first chilling the emulsion on a flaking roll. This is a large revolving drum filled with coolant. The emulsion flows onto the drum in a thin film. It is chilled and scraped off as a soft ribbon or flake. The chilled emulsion is allowed to rest for at least 1 hr to become firm enough to be worked further. It is then pressed through a noodling machine which resembles a large meat grinder. The noodles are allowed to rest for an hour or so and are mixed in a sigma blade dough mixer. The mass is then transferred to a printer which extrudes and cuts it into 5-lb prints. Margarine is also produced in 2.5-lb sheets.

Water chilled pastry margarine is still being produced. In this process, the emulsion is pumped or drained into a bath of crushed ice and water. The chilled emulsion is skimmed off, drained free of excess water, and worked in a kneading machine. The mass is extruded as before into 5-lb prints.

Butter may be incorporated into margarine by melting and mixing it with the margarine emulsion so that they may be chilled together. Solid butter may be blended with solidified margarine in a heavy duty dough or paste mixer and transferred by hand to a printer or filled into large friction top cans. Partially softened butter and margarine may also be forced continuously through an Oakes mixer. The Oakes mixer consists of a disc shaped rotor fitted with pins or teeth which intermesh with a similarly fitted stator. The rotor revolves at high speed and causes intense mixing of the material pumped through it (Fig. 17, 18, and 19).

The major problem involved in blending solid butter and margarine is that if both products are at the same temperature, they will not usually have the same consistency. This will cause the mixture to be lumpy or shotty. Bringing them to the proper temperatures for each to have the same consistency is difficult. It also results in a temperature disparity between the two components of the mixture, making control of the mixing process difficult.

Large margarine producers usually melt the butter, mix it with margarine emulsion, and rechill the blend. Small processors cannot afford the

Courtesy of E. T. Oakes Corp.

FIG. 17.   OAKES MIXER

Rotary positive displacement pump to supply
mixer is mounted at bottom of supporting
frame.

Courtesy of E. T. Oakes Corp.

FIG. 18.   ROTOR WITH STATOR COVERS OF OAKES MIXER

The interior of the covers have pins inserted to mesh with
rotor pins.

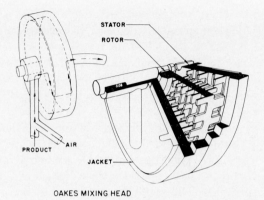

OAKES MIXING HEAD

Courtesy of E. T. Oakes Corp.

Fig. 19. Cutaway Diagram of Assembled
Oakes Mixer Head .

investment required for a complete margarine plant. They are forced to work with a solids mixer and, most often, a hand filled operation.

TABLE GRADE MARGARINE

Oil Formulation

Regular margarine is formulated by blending soybean oil hydrogenated to 2 or 3 different degrees of hardness. This permits the margarine to be spreadable directly out of the refrigerator and to hold together at room temperature.

The physical characteristics of margarine correlate well with SFI measurements of the oil from which it is made. Table 20 gives SFI values for various types of margarine. Values over 30 at 50°F will result in a brittle, nonspreadable margarine at that temperature. An SFI below 28 will definitely produce spreadability. The 28–30 range will be questionable and depend on chilling and working conditions during packaging.

Soft margarines in printed form have SFI values at 50°F from 21 to 24. They must be chilled to a lower temperature in the Votator than is required for regular margarine in order to obtain a print hard enough to wrap properly. The soft margarines were originally formulated by blending hard and soft components, each with a higher iodine value than was used for regular margarine. An increase of five iodine value units for each component would usually be sufficient.

The interest in polyunsaturated oils for nutritional purposes brought about the incorporation of unhydrogenated oils in margarine formulation. The new blends consisted of 50–70% liquid vegetable oil and 30–50%

hydrogenated oil. The hydrogenated component was soybean oil with an
iodine value of about 60 or cottonseed, corn, or safflower oil of equivalent
hardness. The liquid oil depended on which oil was to be promoted:
corn, cottonseed, or safflower. Soybean oil is normally too unstable to be
used at such high levels without being hydrogenated. Some brands of
regular margarine contain 5–10% liquid soybean oil without apparent
flavor problems.

Soft margarine oil blends using beta crystalline hard portions are difficult to handle in chilling. They do not reharden physically if they are de-
formed during printing. This is probably due to their larger crystal size.
Beta-prime hard portions have smaller crystals. Such oils soften on being
worked during printing of the chilled margarine, but the print rehardens
readily after a brief resting period.

TABLE 20

SFI VALUES OF VARIOUS MARGARINE OILS IN THE UNITED STATES

Margarine Type	SFI Values				
	50°F	70°F	80°F	92°F	100°F
Stick (3 component)	28	16	12	2–3	0
80% Liquid oil print	15	11	9	5	2
Cup products	13	8	6	2	0
Liquid oil + 5% hardfat	7	6	6	5.4	4.8
Bakers' regular	27	17	16	12	8
Danish roll-in	29	24	22	16	12
Puff paste					
A-V	25	21	20	16	15
Vegetable	25	24	23	22	21

Source: Wiedermann (1968).

The other end of the temperature scale at which margarine is handled
is also significant. Margarine must melt readily in the mouth with a minimum of waxiness or greasiness. The SFI value at 92°F seems to correlate
well with this requirement. It would be simple if no other factors were
involved. However, the ability of margarine to resist separation into oil
and aqueous phase after prolonged standing at high room temperature is
also related to the SFI value at that temperature.

Low cost margarines with an SFI range of 3.5–5.5 at 92°F will not
soften excessively at room temperature. Such margarine does not have to
be refrigerated. Unfortunately, margarine of this type is waxy in the
mouth and will melt with difficulty on foods which are only warm on being
served.

At one time, the present low cost margarine was the standard of the
industry. Improved formulations and consumer demand for softer products have forced the entire industry to adopt refrigeration for all margarine

products. When the retailer has insufficient refrigerator space, the low cost and sometimes the regular brands of margarine are not held under refrigeration.

Margarine with an SFI value below 3.5 at 92°F will melt well in the mouth. When the SFI value falls below 1.5 with a related value of about 16 at 70°F, the margarine will require refrigeration. It should also be refrigerated in the home between mealtime service, a practice which is often not observed with the firmer brands of margarine.

Soft margarines which are filled in tubs are formulated with larger proportions of liquid oil than can be incorporated into printed margarine. The hardened oil component may be a single oil hydrogenated to an iodine value of 60–65. The margarine oil will contain 75–85% liquid oil. Another soft margarine may be an ordinary margarine oil blend of hard and soft hydrogenated components with 50% liquid oil added. This formulation will probably require 1–2% hardfat as a plasticizing and firming agent.

Two fluid margarines are in limited production. They are basically different from each other in formulation. Pichel (1967) calls for the addition of 0.75–5% hardfat to a salad oil. The basic principal behind the formulation is that a margarine emulsion will not break if it contains a minimum amount of fat crystals. Also, the emulsion will be fluid and pourable if the amount of solid fat is below a definite level. These two requirements set the upper and lower limits of hardfat content given. The SFI value of hardfat in liquid oil changes very little with change in temperature between 40° and 92°F. The use of salad oil as a base ensures that the oil blend will remain fluid and unchanged in the refrigerator.

There also would be some relationship between crystal structure and pourability as is found in fluid shortening. The beta crystal required for fluid shortening seems to be detrimental for fluid margarine, possibly due to the presence of the aqueous phase. Rapeseed hardfat and blends of rapeseed and cottonseed hardfat, which are stable in the beta-prime phase, appear to be preferred for fluid margarine formulation.

The Fricks (1966, 1968) product is prepared by blending regular margarine oil with an equal quantity of liquid vegetable oil. Since the SFI of this blend varies with temperature, there is a temperature range within which the Fricks margarine will remain fluid and pourable. The SFI limits are probably the same as those set forth by Pichel. While the Pichel product would seem to remain stable over a 50° change in temperature (40°–90°F), the Fricks margarine will be semisolid at 40°F and melt at 80°F. The Fricks product can be squeezed from a plastic bottle at refrigerator temperature. However, it could not be classified as fluid under those conditions.

Salt-free margarine is being marketed although the demand for it is limited. The first such product to appear in the United States was distributed in the frozen state to prevent spoilage. Refrigerator stable salt-free margarine has also been produced. It was preserved with benzoic acid. The use of sorbic acid as a preservative was patented by Melnick *et al.* (1964) but its use was not permitted by the Standards at that time. The product was prepared with uncultured milk. Stability of the margarine would have been improved if the pH had been lowered through the use of lactic acid from cultured milk. This was not done, however.

Dietary margarines are available in which the fat content has been reduced to 40%. These products are not covered by Standards. They do not contain milk or other protein as it would not be practical to attempt to prevent microbial spoilage at the water concentration involved. They do contain sorbates and benzoates to prevent spoilage from some organisms, especially molds which grow on fatty emulsions. This does not present as serious a problem as would be obtained from protein decomposition.

Dietary margarines are formulated from oil blends similar to those used for tub margarines. They are also packed in tubs. Some products contain gums and BHA and BHT which are not permitted in standardized margarine. Dietary margarines are not recommended for frying as they contain about 60% water.

Packaging

Most margarines sold for household use are packed as four individually wrapped $1/4$-lb sticks or prints. The maximum legal weight for a margarine package is 1 lb. Each print must be labeled and carry a clear ingredient declaration. It must weigh $1/4$-lb within close tolerances since it is not uncommon for a shopper to break open a 1-lb carton of margarine to purchase 1 stick.

The print former, wrapper, and cartoner is usually a single machine. There are a number of adjustments which can be made to vary its overall performance, depending on the characteristics of the margarine to be packed. The worm screw impellers which feed the margarine noodles to the print mold can be varied in turning rate in relation to the print ejection rate from the mold.

Increasing the impeller rpm increases pressure on the mold and working action on the margarine. This results in softening the print at the time of its formation. It is more dense as it contains less entrapped air pockets. The margarine will also be smoother after it has firmed up on standing. The margarine print is wrapped by forcing it against the paper or foil wrapper. The four sticks are cartoned by pushing them against the carton blank. Too soft a print will collapse under this pressure and the carton will be misshaped and unattractive.

Decreasing the impeller rpm will result in a firmer print which can be wrapped and cartoned easily. It may also result in a crumbly stick full of unattractive air pockets. The correct impeller speed in relation to the print forming rate lies between the two extremes. It will differ for various margarine oil formulations and the manner in which they are chilled.

The molding chamber size can be adjusted to compensate for density of the margarine and air pocket distribution. This is essential in order to maintain legal weight requirements.

Machinery for wrapping whipped margarine in print form at six sticks per pound must be designed to handle the prints gently. The machine should pick up the stick in a partially folded wrapper and complete the folding without applying pressure to margarine. The sticks are then assembled in a group of six and inserted into a prefolded carton.

Print wrapping material may be either parchment paper or paper-foil laminate. Foil is not only more attractive, it is also impermeable to moisture and air. Parchment paper is permeable and permits moisture loss from the margarine surface. This loss causes the emulsion to break on the surface. The moisture then appears darker in color but the interior remains light. The surface oil is also exposed to oxygen in the air and tends to become rancid. Prints of nationally distributed premium brands of margarine are usually foil wrapped. Low cost margarine is wrapped in parchment paper as an economy measure.

Economy is also practiced in the use of wax or plastic impregnated paperboard cartons for low cost margarine. They are usually tuck-folded. Premium brand cartons are either foil laminated board or are overwrapped with foil laminated paper. They are also sealed and supposedly tamperproof.

Prints of $1/_2$- and 1-lb sizes are normally wrapped in parchment paper and are not cartoned. The 1-lb size is often packed for institution use.

Soft margarines are filled into 8-oz tubs and packed 2 in a carton. High-density polyethylene is the most frequently used material. Plastic coated aluminum cups are also used. The coating is required to prevent corrosion of aluminum by salt in the margarine. Polyethylene does not corrode but it is oxygen permeable. This will ultimately cause the margarine to become rancid. The necessity for refrigerating soft margarine to prevent emulsion separation also retards development of rancidity to a practical degree.

Fluid margarines are packaged in polyethylene bottles. These bottles are highly permeable to oxygen with a great risk of the product becoming rancid even under refrigeration.

BAKERY MARGARINES

Although bakery margarines can be considered as shortening in emulsified form, they are decidedly not emulsified shortenings. The level of monoglyceride permitted under the Standards is insufficient to function as an emulsifier in cake and icing work. Lecithin is often omitted as bakery margarines are not used for frying. A few products are being marketed which are essentially margarines with a high monoglyceride content. These are sold under brand names and carefully avoid the designation "margarine."

As with bakery shortenings, margarines are designed for specific uses. The major difference between the various margarines must be in fat composition. The only other variations permitted under the Standards relate to addition or omission of color, choice of flavor, salt level, and the like.

Margarines are rarely aerated as are shortenings. The emulsification with milk or water has the same ability to provide opacity and brightness by reflection as does whipping with air. Margarine is more dense than shortening. It too is packed in 50-lb slip top cans and cubes but the dimension of these containers is smaller. Open-end 55-gal. steel drums hold about 420 lb of margarine as compared to 380 lb for shortening. Pastry margarines are extruded and cut in 5-lb blocks and 2.5-lb sheets. They are usually wrapped in parchment paper and packed in corrugated paperboard boxes.

One type of bakery margarine is simply table margarine filled into a large container. A shortening type working B-unit is used for finishing instead of the static B-unit. This margarine is used primarily for preparation of cream icings. It is preferred to shortening for this use as margarine has flavor, color, and good melting characteristics. The wide plastic range shortening which would be used otherwise produces a bland icing with a slightly greasy aftertaste.

Ordinary bakery margarine does have a wide plastic range. Like its shortening counterpart, it is, in effect, a general purpose margarine. It is used in baking cookies, pound cakes, and pastries. Most of these margarines are made from the same base oil type as is used for table margarine with 4–8% hardfat added as a plasticizer.

Flavor and color are the main distinctions between margarine and shortening for cake and cookie baking. The flavor is usually strong and preferably contains butyric acid. Diacetyl is quite volatile and is usually lost during baking. Butyric acid does not bake out readily and contributes some flavor to the finished baked products.

Bakery margarines contain about $1/2$ oz salt per pound of product. This should be taken into account when switching a cake or cookie formula from shortening to margarine. Butter has about $1/4$ oz or less salt per pound.   Butter cookies baked with margarine may be too salty if the cookie dough formula is balanced for butter use without added salt.   If none has been added, there will be none to take away.

The general purpose product is also used as a roll-in margarine for Danish pastry preparation.   Danish pastry is made by rolling a prepared dough, usually yeast-raised, into a sheet about $1/_2$ in. thick.   The margarine is spotted on the dough by scattering approximately 1-in. diam pieces of it over $1/_3$ of the dough surface.   The dough is folded to cover the margarine and again spotted with additional margarine.   This is re-folded to give two layers of margarine between three layers of dough. The dough is then chilled for several hours.

After being chilled, the dough is very stiff but pliable.   The margarine should be the same.   The dough is rolled, folded, and rerolled several times.   If the margarine is too soft, it will squirt out of the dough on being rolled.   If the margarine is too hard and brittle, the margarine lumps will tear the dough on being rolled, causing it to lose flakiness.   Baking of pastry dough in which hard lumps of margarine persist after rolling will contain voids where the margarine lumps have melted away.

It is difficult to produce an all-purpose margarine which will have the correct workability at refrigerator temperature for use in Danish pastry. The margarine oil formulation is only part of the answer.   The margarine must be chilled properly.   It must be filled at such temperature that it forms a slight mound in the container.   Too hot or too cold a fill will result in too hard and brittle a margarine.   Proper tempering is also important. The margarine should be held for about 48 hr at its filling temperature to attain optimum low temperature plasticity.

Blends of tallow and soybean oil, with or without hydrogenation and with small amounts of added hardfats, are used for animal-vegetable (A-V) general purpose margarines.   The products are chilled, filled, and tempered in the same manner as all-vegetable margarines.

The Standards will not permit the use of antioxidants in margarine. Unhydrogenated meat fat margarines must be refrigerated to obtain stability against oxidative rancidity.   Bakery margarines are not tasted. The baker is aware of odor.   Reverted odors in margarine are easily masked by the high level of artificial flavor normally added.   The baking process is also effective in deodorizing oils.   The heat and evolution of steam seem to break down and boil off compounds which result in the formation of reverted flavor.   Margarines of this type are not suitable for preparation of icings.

Blends of butter and margarine are desired by specialty chefs and bake-shops.   The flavor effects and melting characteristics obtained seem to be worth the premium price required for such products.   Butter levels of

25, 35, and 50% are most common. Butter oil and margarine oil are quite different in fatty acid composition. Butter oil contains about 25% fatty acids with chain lengths of less than 16 carbon atoms. On blending any dissimilar oils, the resulting mixture is always softer than either component alone. This is probably due to intersolubility of the various triglycerides. Margarine and butter blends, therefore, require the addition of a few percent hardfat flakes to compensate for the loss in firmness. Otherwise the finished margarine would tend to be soupy.

Special roll-in margarines are designed entirely for preparation of Danish pastry. They are usually formulated from unhydrogenated tallow and soybean oil in approximately equal amounts and plasticized with about 8% hardfat. This type of margarine differs from all-purpose A-V margarine in that it is chilled on an external flaking roll while the general purpose product is chilled in a Votator system. The mechanical action resulting from working the margarine flakes into the final print gives the product a characteristic plasticity suitable for rolling into pastry dough.

Roll-in margarine is packed in 5-lb prints as the average formula for pastry requires two 2.5-lb pieces of margarine for the standard size batch of dough. This size is easy for the baker to roll by hand. Pastry rolling machines are available. Some are designed to handle the same size batch.

Puff paste is, in a sense, a type of roll-in "margarine." However, it is defined as a shortening and is not covered by the margarine standards. Puff paste contains 90% fat. It is usually churned with water rather than milk. This would not have been acceptable under the Standards prior to 1966, but is permitted today.

Puff paste is used exclusively for making puff pastry. Puff pastry is an expanded, flaky, baked product best known for making turnovers, patty shells, and Napoleons. It is prepared by rolling the puff paste into a tough but pliable dough in the same manner as with Danish pastry. Roll-in margarine is soft and soaks into the layers of dough to some extent so that it becomes flaky but does not puff. Puff paste is firm and waxy. It insulates the dough layers and causes the dough to "spring" or expand into puffed form.

A so-called "blitz" method has been developed for puff pastry production. The dough is put in an ordinary bakery mixer with lumps of puff paste and mixed until the paste is distributed thoroughly in finely divided particles. It is then rolled out and cut into individual pieces for baking. The pastry is not as good as the hand rolled product, but is less costly to produce. The amount of hand labor required for pastry production by the old method had almost priced puff pastry off the market.

A-V puff paste is formulated from blends of unhydrogenated tallow and soybean oil in approximately equal amounts and about 15% hardfat.

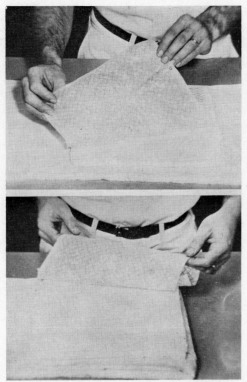

Courtesy of Swift & Co.

FIG. 20. PREPARATION OF PUFF PASTRY

Top—shows placement of first sheet of puff
paste; bottom—shows placement of second sheet.
Last flap of dough is then folded over the
second sheet for rolling.

The oil differs from roll-in margarine oil only in that a higher level of
hardfat is used. This makes an extremely firm and waxy puff paste.

A quick evaluation for potential puff paste performance is to pick up a
small wad of it and knead it between the fingers and palm of the hand.
A good paste will act like modeling clay. It will change shape without
feeling brittle or lumpy. It will work without feeling sticky or falling
apart.

Vegetable puff paste is made by blending about 25% cottonseed oil
hardfat with unhydrogenated vegetable oil, usually soybean oil. The
product does not produce as good a pastry as one made from an A-V
formula. Vegetable pastry springs well but usually unevenly. The
kneading test shows the vegetable paste to be somewhat sticky. This

might explain its poor performance in that the paste may penetrate the dough layer to some extent and especially unevenly.

An improved vegetable paste could probably be made by incorporating some triglycerides of intermediate hardness. Meat fat paste contains liquid oil, partially saturated triglycerides, and fully saturated triglycerides. Vegetable paste contains only liquid oil and fully saturated fats.

Puff pastes are made by chilling the oil and water emulsion on the flaking roll or in ice water. They are packaged in 5-lb prints or 2.5-lb sheets. The dough batch is sized accordingly as with roll-in pastry. Sheeted paste was developed to eliminate the time required for spotting paste from prints. Figure 20 illustrates how the sheets are layed on the rolled-out dough. The puff paste sandwich is then rolled, folded, rerolled, and refolded several times before being cut into sections for baking.

Although bakery margarines are formulated with salt and sodium benzoate, mix-handling can cause mold growth. Cans and drums of margarine have a head space which can sweat on being placed in refrigerated storage. The condensate will dilute the salt and benzoate where it contacts the margarine surface to a point where these compounds can no longer protect the product from spoilage. Parchment paper liners are used to cover the margarine surface before the lid is placed on the container. This ordinarily prevents headspace condensate from reaching the margarine surface.

Margarine print machines have water jacketed molds and extrusion heads. Leakage of water from lines leading to these jackets can also dilute margarine preservatives and cause spoilage.

BIBLIOGRAPHY

ANDERSON, A. J. C., and WILLIAMS, P. N. 1965. Margarine, 2nd Edition. Pergamon Press, New York.

ANON. 1962. Sheeted margarine assures uniform flakiness of pastry. Food Process. 23, No. 4, 66–67.

ANON. 1965. Margarine in a bottle. Food Process, 26, No. 3, 44.

BOLANOWSKI, J. P., and ZIEMBA, J. V. 1966. For higher process efficiency: Continuous emulsification. Food Eng. 38, No. 11, 86-91.

FRICKS, W. E. 1966. Plasticizing apparatus for producing uniformly emulsified margarine. U.S. Pat. 3,269,012. Aug. 30.

FRICKS, W. E. 1968. Method for producing pourable refrigerated margarine. U.S. Pat. 3,397,998. Aug. 20.

MELNICK, D. et al. 1964. Unsalted margarine. Can. Pat. 698,683. Nov. 20.

PICHEL, M. J. 1967. Fluid margarine emulsion stabilized by hardfat. U.S. Pat. 3,338,720. Aug. 29.

TODD, P. H., JR. 1964. Vegetable base food coloring for oleomargarine and the like. U.S. Pat. 3,162,538. Dec. 22.

WIEDERMANN, L. H. 1968. Margarine oil formulation and control. J. Am. Oil Chemists' Soc. 45, 515A, 520A–522A, 560A.

CHAPTER 10

Mayonnaise and Salad Dressings

MAYONNAISE

Standards of Identity

Mayonnaise is defined under U.S. FDA Standards of Identity as an emulsified semisolid food prepared from edible vegetable oil, acetic or citric acid, and egg yolk. Optional ingredients permitted include salt, natural sweeteners, spices or spice oils, monosodium glutamate, and any suitable harmless flavor from natural sources. These various ingredients have a number of limitations imposed on them.

The oil level must be not less than 65% by weight of the mayonnaise. The oil may contain 0.125% oxystearin to inhibit fatty crystal formation.

The egg yolk may be in the form of separated yolk or whole egg. Egg yolks may be liquid, frozen, or dried. They may contain added liquid or frozen egg white. From a practical standpoint, this means that commercial eggs which are separated on breaking into yolk and white may be recombined in any proportion provided some egg yolk component is used. Of course, the oil would not form a stable emulsion in the complete absence of egg yolk.

Vinegar is the source of acetic acid. Mayonnaise must contain not less than 2.5% acetic acid by weight. A lower concentration will not protect the mayonnaise from spoilage. Citric acid may be used as a substitute in an amount not exceeding 25% of the total weight of acid. Citric acid may be used as the only acid at a minimum level of 2.5% when incorporated as lemon or lime juice.

The spices, spice oils, or other seasonings permitted must not impart a color simulating that provided by egg yolk. This eliminates the use of saffron or turmeric. Mustard and paprika are permitted. Mustard does contribute a yellowish cast but is highly flavored. It cannot be added in sufficient quantity to simulate a high egg content without being overbearing in heat and flavor. Oleoresin paprika may impart a slight yellowish color at a low level, about 0.01%. Mayonnaise containing higher levels of paprika tends to be too red in hue. The paprika level is therefore self-limiting.

Disodium and calcium disodium EDTA salts are permitted as metal scavengers at levels up to 75 ppm. They are claimed to protect the oil from oxidizing or reverting in flavor and to protect the mayonnaise from loss in color.

Mayonnaise may be whipped with and packed under inert gas, i.e., nitrogen or carbon dioxide.

Formulation and Ingredients

Formulation.—Most commercial mayonnaise falls within the formula limits given in Table 21. Each ingredient has a specific function. The quantity of each is carefully worked out to achieve the characteristics desired by the manufacturer. The proportions of oil and egg are balanced to obtain body, viscosity, and texture. These are also related to the equipment available and to the manner in which the equipment is operated. Salt, sugar, vinegar, and spicing are balanced to give a smooth, rich flavor. However, the character of the emulsion is also involved in flavor percep-

TABLE 21

MAYONNAISE COMPOSITION

Ingredient	Weight %
Salad oil	77.0–82.0
Fluid egg yolk[1]	5.3–5.8
Vinegar (100 gr)	2.8–4.5
Salt	1.2–1.8
Sugar	1.0–2.5
Mustard flour[2]	0.2–0.8
Oleoresin paprika[3]	
Garlic, onion, spices[2]	
Water to make 100%	

[1] Egg solids, 43%. May substitute whole or fortified egg, fluid or dry, on a total solids basis.
[2] Spice oils, oleoresins may be substituted.
[3] Optional where characteristic color is desired.

tion. A tight emulsion results in mild flavor. A weak emulsion can emphasize the sweetness, tartness, and saltiness making a poorly balanced flavor especially apparent.

Oil.—Salad oil is used in making mayonnaise. The source may be winterized cottonseed, unhydrogenated or hydrogenated, and winterized soybean, safflower, corn, or olive oil. Unless the manufacturer wishes to make special claims concerning flavor or nutrition, unhydrogenated soybean oil is the one normally selected. It is usually the least costly oil. An oil which solidifies at refrigerator temperature is avoided as the mayonnaise emulsion will break as soon as the oil begins to crystallize. The emulsion will also break if the aqueous phase freezes.

The oil used is normally deodorized. Gourmet mayonnaise is sometimes made with virgin olive oil to obtain a unique flavor. As little as 10% olive oil will have a noticeable effect.

Mayonnaise is an oil-in-water emulsion. Droplets of oil are dispersed in a continuous aqueous phase. The rigidity of the emulsion depends partly on the size of the oil droplets and how tightly they are packed. The more oil that is dispersed in the emulsion, the stiffer it will be.

The consuming public has become accustomed to a definite viscosity range for commercial mayonnaise. This has resulted in the 77–82% oil content limitations given in Table 21. The Standards permit as little as 65% oil but the resulting product would be considered as too thin for the current market. Oil levels of 80–84% make a thick or heavy-bodied mayonnaise. Such product is preferred for institutional use as it does not soak into bread in sandwiches or soften and flow over salads. The housewife, however, finds thick mayonnaise too dry and even rubbery for her taste.

More than 84% oil will result in overloading the system. The droplets will be packed too tightly with too thin a wall between them. Mechanical shock could easily cause the oil droplets to coalesce and break the emulsion.

Egg.—The amount and type of egg solids has an effect on emulsion viscosity and strength. Egg yolk contributes the major emulsifying compounds to the mayonnaise formula. Synthetic emulsifiers are not permitted under the Standards. Joffe (1942) lists the egg yolk emulsifiers as lecithoproteins, phospholipids, and cholesterol. Egg white protein aids in emulsification by forming a solid gel structure on being coagulated by the acid component (Flückiger 1966). The more emulsifier and solid matter that can be dispersed in the aqueous phase in colloidal form, the more rigid this phase becomes. Conversely, there is a lower limit for egg content below which a stable emulsion cannot be formed.

When eggs are broken on a commercial basis, they are separated into yolk and white. Commercial yolk contains 43% solids and constitutes 40% of the egg. The actual yolk averages only about 35% of the egg. The balance is egg white which cannot be separated without difficulty. Yolk and white can be recombined in any desired proportion. Whole egg contains about 26% solids. When the solids content is greater than this amount through the use of more yolk than normal, the combination is known as fortified egg. Egg with 33% solids is a common level of fortification.

Several brands of mayonnaise use fortified egg. Substitution of fortified egg for yolk has little effect on measurable viscosity of the mayonnaise if the total egg solids content of the formulation is held constant. The texture of fortified egg mayonnaise differs from that made from yolk alone. The presence of additional egg white results in a meringue-like fluffiness not obtainable without it.

Eggs for mayonanise may be fresh, frozen, or dried. Fresh broken eggs usually make a weak-bodied mayonnaise although the product may stiffen somewhat after storage (Kilgore 1935). Yolk proteins begin to gel on being frozen, reaching a maximum gelation point in 2 to 4 weeks. This phenomenon is accelerated by deep freezing the egg to —20°F for 72 hr, and transferring the product to 0°F for the balance of the storage period.

Yolk by itself will gel irreversibly on freezing, becoming indispersible and useless for mayonnaise production. It will resemble gum rubber in texture. Dilution of the yolk with 10% salt, 10% sugar, or egg white will permit partial gelation. The thawed egg will be thick but dispersible (Reynolds and Harris 1932). Mayonnaise is usually prepared from egg thickened by freezing. The mayonnaise will also be thick and creamy. Yolk frozen with 10% salt added is used most frequently as it is resistant to microbial spoilage during thawing.

Dried egg is useful for mayonnaise production. It will disperse readily in the aqueous portion of the mayonnaise formulation. The resulting product is thicker than one obtained from frozen egg when prepared with the same total egg solids content. Approximately identical mayonnaise viscosity can be obtained from dried egg when it is used at a solids level of 95% of that of frozen fluid egg (Slosberg 1968).

Egg yolk is the primary source of yellow color in mayonnaise. No other coloring matter is permitted. Oil contributes very little yellowness although chlorophyll in the oil will contribute a greenish cast. This is especially noticeable in safflower and olive oil mayonnaise.

Egg varies considerably in yellow color content between producing areas. Color is measured by the National Egg Products Association (NEPA) method (Finberg 1955). West coast egg is generally lighter in color than midwest and eastern egg. In addition, ground-raised chickens in the midwest and east lay dark eggs in the spring and early summer. West coast wire-raised chickens are fed prepared feed throughout the year. Color variations in eggs from these chickens would be due to feed ingredient variations. For example, corn in the feed gives darker yolk than sorghum. The relative market price of corn and sorghum determines the yolk color and consequently the mayonnaise color obtained.

Pasteurization of egg products was ordered by the U.S. FDA to eliminate *Salmonella* contamination. It was found that pasteurization of salted yolk or salted fortified egg did not adversely affect its performance in mayonnaise emulsification.

Acids.—Acid is the main preservative against microbial spoilage in mayonnaise. Distilled vinegar is the most commonly used acid in mayonnaise preparation. Vinegar strength is measured by "grain," 100 gr equaling 10% acetic acid. Industrial vinegar is usually available as 100 and 120 gr. Distilled vinegar is the lowest in cost of all the vinegars and is also less costly than citric acid. Lemon juice is added to some brands of mayonnaise, partially for flavor effect, partially for promotional reasons. Gourmet mayonnaise may be made entirely from lemon or lime juice but this would be too costly for ordinary commercial products.

Not all distilled vinegars are equivalent to each other in flavor. They will vary between producers and contain various levels of ethyl acetate. This results from the reaction between the ethyl alcohol starting material and acetic acid formed by oxidation of the alcohol. Acetaldehyde, an oxidation intermediate, may be present as may other minor components. These all affect vinegar flavor and the flavor of mayonnaise prepared from it. Vinegar flavor can be altered by filtration through activated charcoal.

Cider, malt, and wine vinegars are more costly than distilled vinegar. They have unique flavors which contribute character to mayonnaise. A small amount of specialty vinegar may be desirable for gourmet type products. A large proportion of such vinegar would probably result in sufficient flavor for the mayonnaise to be considered as spoiled. These specialty vinegars are dark in color. Mayonnaise prepared from them may be too dark. Charcoal filtration will bleach the color but may also remove some of the desirable flavor notes.

Mustard.—More conflicting opinions have been expressed concerning the effect of mustard flour on mayonnaise than about any other ingredient in the formulation. Kilgore (1932) considered mustard flour to possess important emulsifying properties. Cumming (1964) has observed that oil of mustard could be used in place of mustard flour without affecting emulsion strength. It would seem that the activity of mustard flour as an emulsifier would depend on the type of mustard, the balance of various ingredients in the specific formula used, and the process by which the mayonnaise is prepared (Kilgore 1933).

In some evaluations of mustard flour, the balance between oil and egg content may have been on the borderline of failure of attaining an emulsion. The presence of mustard flour may have been critical for obtaining emulsification. An increase in the amount of egg might have lessened the importance of mustard in such a case.

Much of the early work on the effectiveness of mustard flour in emulsion systems was done on inefficient equipment. Modern colloid mills have probably offset the reliance on emulsifier performance of mustard as a major factor in mayonnaise production.

Mustard flour has been found to affect the character of mayonnaise in storage depending on preconditioning of the flour. Mustard seed contains protein, starch, neutral oil (triglycerides), and glucosides among many other components. Oil of mustard, or allyl isothiocyanate, is extremely pungent in free form but is flavorless in combined form as the glucoside. Wetting the mustard flour with water activates glucosidase enzymes which release the oil of mustard.

Other compounds are also apparently activated by water (Kilgore 1934). These materials cause progressive weakening of the mayonnaise emulsion and rapid flavor deterioration. Preparing mustard with vinegar in place of water seems to eliminate the compounds which cause this breakdown. Since most mayonnaise is prepared by suspending mustard flour, if used, in the egg-oil mix and adding preblended vinegar and water, activation of these undesirable materials is probably an academic rather than a practical problem.

Mustard seed is available in two varieties, yellow and brown. They have also been called white and black. The two varieties are blended in various proportions by the supplier in order to achieve a desired balance between flavor and pungency. Yellow mustard is hot to taste but odorless. Brown mustard has a sharp odor (Bice 1965). Selection of mustard seed for use in mayonnaise is done with great care. Weed seed, often wild rapeseed, is a common contaminant. Such seeds have a black seed coat which appears as black specks when ground in with the mustard flour. These specks are especially noticeable against the light colored background of mayonnaise. If the mustard seed has not been properly cleaned or the flour not properly screened, the specks will be too large in both size and quantity.

The mustard flour components which are responsible for heat and flavor change in a relatively short time, perhaps two weeks after the mayonnaise is prepared. Heat is no longer apparent as such. A new egg-like flavor develops (Bice 1965).

Some mayonnaise manufacturers use oil of mustard in place of mustard flour. The Standards require that the oil be obtained from mustard seed. Synthetic allyl isothiocyanate is not permitted. Mustard flour contains 0.5–1.0% available oil of mustard depending on the seed blend used. Replacement of flour with oil has been recommended by Cumming (1964). The amount of oil suggested is about $1/_3$ of that obtained from flour as the oil is supposedly more pungent in the free state.

Mustard oil has certain advantages over flour. The oil is speck-free. It retains its original pungency and flavor for a longer time, perhaps for three months. The oil has no protein or starch component to cause potential emulsion breakdown.

The oil is extremely hazardous to use! It is a powerful irritant. It will burn and blister skin on contact. Its vapors will burn the eyes and nasal passages. Diluting oil of mustard in salad oil is perhaps the best way to handle it. If the concentration in salad oil is 0.25–0.5%, the solution can be added to the mayonnaise formulation at the same level as would be used for mustard flour.

Mustard flour does contribute color to the mayonnaise emulsion where mustard oil does not. Consumer acceptance of color in mayonnaise is subject to regional preference. A decision to use the flour or oil should take finished product color into account.

Paprika.—Paprika is normally used in mayonnaise as the oleoresin. This is the oil which results from solvent extraction of dried paprika pods. The oleoresin is rated by color value with 60,000 and 120,000 unit products being commonly available. The units are based on matching diluted oleoresin with special color standards.

Oleoresin paprika has a characteristic odor and either a mild or hot flavor depending on its source. The mildest product is used for mayonnaise. The amount required is small, in the order of 1 oz per 10,000 lb finished mayonnaise. The coloring effect is quite apparent. The flavor of paprika is too mild to be noticeable at this level.

Seasoning.—Salt and sugar are added to mayonnaise at what appears to be a low level. Saltiness and sweetness, however, are functions of concentration in the aqueous phase of an emulsion system. With mayonnaise containing 80% oil, the salt and sugar are dissolved in less than 20% water and are, therefore, relatively concentrated.

Spices, herbs, and natural oils obtained from them are also used as seasoning. The most popular spice, mustard, is usually considered as a separate subject since it is conceived of as being a necessary ingredient in the preparation of mayonnaise. This is not the case, however. The Standards of Identity include mustard as an optional ingredient. Mustard flour is not required as an emulsifier. Black pepper oleorsein or oil can replace mustard as a source of heat (Cumming 1964).

Garlic powder is frequently used in mayonnaise. Some regionally distributed brands contain noticeable quantities. Others are more subtly flavored. The characteristic flavor of garlic seems to be quite stable on long storage.

Onion is milder in flavor than garlic. Unfortunately it is also much less stable. Mayonnaise will not retain a characteristic onion flavor for more than a few days. The residual flavor may have some beneficial influence on other flavors in the product, however.

Other spices are not often used in mayonnaise. Tarragon, clove or allspice, and cinnamon oils have been recommended. Lemon peel oil may be added, but it seems preferable to add lemon juice with a high natural oil content.

Inert Gas.—Carbon dioxide or nitrogen is permitted as an inert gas for packaging of mayonnaise. Nitrogen seems to be preferred. The gas is introduced into the processing system just ahead of the colloid mill.

The specific gravity of degassed mayonnaise is about 0.94. The volume of inert gas added is usually adjusted to give a finished product specific gravity of 0.88–0.90 at the time of manufacture. A lower gravity is not practical. Mayonnaise loses gas on standing and especially during shipment to market. Most mayonnaise will hold sufficient gas to retain a 0.90–0.92 sp gr after normal shipping and storage. Excessive gas injection to attain too low a specific gravity will result in sufficient product volume loss in the package to be objectionable to the customer.

Mayonnaise is sold by volume rather than by weight. The lowest practical specific gravity would result in the highest yield of package units per unit of formula weight.

Equipment and Process

Fundamentals.—Mayonnaise is a difficult emulsion to prepare. An emulsion tends to form with the major component in the continuous phase and the minor component in the dispersed phase. Oil is the major phase in mayonnaise. It is forced to exist in the dispersed phase. When a mayonnaise emulsion breaks, it is in effect a reversal of the emulsion to a stable form where the oil becomes continuous and the aqueous portion discontinuous, although not dispersed.

All mayonnaise equipment should be fabricated from stainless steel. Vinegar will corrode ordinary steel and aluminum. Equipment for production of mayonnaise consists of some form of intensive mixer to bring about dispersal of the oil into fine droplets. Process details vary with the equipment used. Most processes are batch type. Some have been modified into continuous systems by preblending the ingredients into two or more components which are then injected into a mixing unit by proportioning pumps.

Filling equipment is an integral part of any mayonnaise system. The emulsion is fluid when first prepared. It should set up into a semisolid gel after a specific length of time. If the emulsion is disturbed after the first gelation, it will become too soft, even though it will regel on further standing. The time required for gelling depends on many interrelated factors in formulation, equipment, and procedure. The filling machinery must be able to package the mayonnaise fast enough for the emulsion to set up in the jar but no sooner than that.

To a great extent, mayonnaise production is an art. The effect of the many factors involved are understood to some degree on an individual basis. The interrelationships concerning variability of these factors are not always clear from a strictly scientific standpoint. Operator experience is important for success. When this is lacking, it is easy to blame failure on the eggs.

Eggs are the most complex ingredient and consequently the least understood. Egg quality can only be evaluated by performance testing at present. Low egg solid content can cause failure which is correctable by increasing the amount of egg material proportionately. A good operator can correct for other egg performance variables by slight modification in the process depending on his careful observation and experience.

Mayonnaise emulsion is easiest to prepare when the fluid ingredients are chilled. Gray and Maier recommend 60°–70°F for product made in small mixers (Mattil 1964). Ingredient temperature should be roughly 50°–60°F when a high-speed colloid mill is used. Temperature of the emulsion will rise about 10°F during milling. Emulsion failure will normally result from an output temperature above 75°F. Some mill heads are jacketed for circulation of coolant. This helps, but heat transfer rates are not high enough to maintain a sufficiently low temperature without precooling feed-stock to the mill.

Planetary Mixers.—The simplest equipment by which mayonnaise can be prepared is an ordinary planetary mixer equipped with a paddle. The Hobart mixer is of this type and is used by gourmet restaurants for preparation of specialty mayonnaises. Similar batch mixers of larger capacity were once used for manufacture of commercial mayonnaise. They have been replaced by the colloid mill and other continuous flow emulsifying mixers.

The classic technique recommended for processing mayonnaise in the planetary mixer is to mix the egg, preferably yolk only, with the dry ingredients. Vinegar and oil are then added slowly in separate streams but at the same time. The oil addition is supposed to be completed slightly before the last of the vinegar is poured in. The resulting mayonnaise is often soft unless a large proportion of egg yolk is used. Failure can result if the oil is added at too rapid a rate. The presence of large oil masses tends to cause the smaller droplets of oil to coalesce, breaking the emulsion.

An improved technique is to add much of the oil to the egg-dry ingredient mix first. The vinegar is added afterwards. Whole egg may be used with the oil-first method as the resulting mayonnaise will have a heavier body than when oil and vinegar are added simultaneously. The oil should be added slowly at first. It can be poured in more rapidly after the mass begins to thicken. Oil should not be allowed to puddle. If it does, however, further oil addition can be stopped temporarily until the free oil is completely dispersed. The advantage in the oil-first method is that the oil will disperse ultimately. Where vinegar and oil are added together, it is frequently impossible to correct for errors in rate of oil addition. Once all of the oil is dispersed in the mix, the flow rate of vinegar addition is not too critical. Flooding of the mayonnaise mix with vinegar should, of course, be avoided.

Viscosity of mayonnaise prepared by the oil-first technique can be controlled within limits set by the formulation. Maximum viscosity is obtained by adding all of the oil to the egg before any vinegar is added (Brown 1949). The viscosity can be lowered by occasional addition of small portions of vinegar while the oil is being added. Operator experience is necessary to judge the relationship between interim viscosity of the egg-oil mix and final thickness of the mayonnaise. When the mixture becomes too thick during the process, a small amount of vinegar is added to thin it out. Emulsion failure is rare with this method as high viscosity of the egg-oil combination aids in dispersing further oil additions. The oil droplets are smaller because of the high viscosity. A high viscosity mayonnaise emulsion is also the result of closely packed small oil droplets.

When oil and vinegar are added simultaneously throughout the entire period of premix makeup, the mix viscosity is too low to be useful in dispersing oil properly. Oil dispersed in this procedure must depend entirely on mixer action. The oil droplets are usually larger than with the oil-first method and the mayonnaise is consequently less viscous.

Dixie-Charlotte System.—The Dixie-Charlotte system is probably the one most widely used for mayonnaise and salad dressing manufacture. The system is available in a number of sizes with capacities ranging from 15 to 200 gal. per batch, yielding 60–1200 gal. of finished mayonnaise per hour. The equipment is arranged on a continuous batch basis and can be modified to be operated as a truly continuous system. The broad range of capacities and the ability to produce a single batch or a continuous series of batches of mayonnaise has led to the popularity of this equipment.

The system consists of two Dixie mixers and a Charlotte colloid mill connected by appropriate piping, valves, and a rotary positive displacement pump (Fig. 21). It is arranged so that one mixer is feeding premix to the mill while another batch of premix is being prepared in the other mixer. The Dixie mixer is a deep circular tank fitted with three turbine mixers mounted side-by-side on a horizontal shaft close to the tank bottom. The shaft is turned by a variable speed motor. This mixer will prepare mayonnaise without further processing. The emulsion will be similar to one made with a Hobart mixer. The creamy texture of modern mayonnaise is obtained by pumping this relatively soft, coarse emulsion through the Charlotte mill.

Preparation of mayonnaise premix in the Dixie mixer is started by adding mayonnaise from a previous run to a level which will reach the mixer shaft. This gives the turbine blades a heavy material to work against while shearing the oil into fine droplets. This is no problem with a continuous series of batches. The premix is merely emptied to the shaft line, leaving sufficient "seed" for the next batch.

Egg and dry ingredients are added to the seed in the mixer with low-speed agitation. Oil and vinegar are then pumped or fed by gravity from supply tanks into the mixer. Agitation speed is usually increased periodically as the level of the premix rises. Mixing at too high a speed when the mix level is low will literally throw product out of the tank. A

Courtesy of Chemicolloid Laboratories, Inc.

Fig.   21.   Dixie-Charlotte   Mayonnaise
System

Rotary positive displacement pump feeds emulsion premix from Dixie mixer tank to Charlotte mill.

relatively full mix tank, however, contains too large a volume to be effectively mixed at lower speeds, especially with thick emulsions.

The premix is pumped through the Charlotte mill as soon as it is made. Timing should be such that the alternate premix tank should have just finished being emptied. This is critical with many mayonnaise formulations as the premix can gel in the Dixie mixer. The longer the premix is held in the mix tank prior to milling, the softer the final product will be.

Fortified egg formulations seem to be more resistant to damage by delay in milling. Premix emulsions prepared from fortified egg are usually thinner than those made from yolk alone. They may take longer to gel. A more plausible explanation is that fortified egg emulsion thickens during milling from dispersion of acid-coagulated egg white as well as from dispersion of oil. This may also account for its meringue-like whipped texture.

Premix viscosity is controllable in the Dixie mixer in the same manner as it can be controlled in the Hobart mixer. Oil is added to the mixture of egg, dry ingredients, and seed until a predetermined viscosity is reached. This is usually judged visually by the mixer operator. Vinegar flow is then started, to run concurrently with the balance of the oil. Oil and vinegar flow rates are timed by using plug cocks or valves to throttle the flow.

Premix viscosity is a determining factor in gelation time of the milled mayonnaise. If it is too high, the mayonnaise will gel in the filler hopper and be permanently softened during the filling operation. A still higher premix viscosity may result in the emulsion breaking during milling. Too low a premix viscosity will result in too low a finished product viscosity.

Control of premix viscosity is desirable in order to compensate for variations in other factors. Such control will not completely compensate for poorly processed egg, for example. It will, however, reduce variability in the final product. It may even allow for preparation of acceptable mayonnaise where lack of control could cause emulsion failure.

Weak and broken emulsions can be reworked by adding them back to the Dixie mixer as seed or in the early stages of premix preparation. The amount of such material should be limited to perhaps 10% of the total weight. Mayonnaise containing rework should be placed in high turnover rate market areas such as the institutional trade. This mayonnaise is not necessarily defective, but it may have a reduced shelf-life.

The Charlotte mill consists of a horizontally mounted, truncated-conic, fluted rotor inserted into a close fitting, fluted and jacketed stator. The spacing between rotor and stator is adjustable. The adjustment wheel which moves the rotor shaft in or out of the stator has a series of holes drilled around it just inside its periphery. Each hole is placed so that the turning of the wheel moves the rotor one mil per hole. A lock pin is inserted in the top hole to prevent unwanted movement of the adjustment wheel. All sizes of the Charlotte mill are designed in this manner.

Each size of mill rotates at a fixed speed of 3600 rpm. Peripheral speed of the rotor increases with mill diameter. The working action of the various sized mills is a function of both peripheral speed and throughput rate of the emulsion. The Waukesha pump which feeds the mill has a variable speed drive. The range of pump capacities is selected to make each size mill system equivalent to all other sizes in working action.

Milling is most effective with the smallest mill opening which will not result in a broken emulsion. This will depend on product formulation and on throughput rate required for the filling equipment. The correct mill opening is found by trial and error. Finished mayonnaise viscosity will increase slowly with narrower mill openings.   However, viscosity will increase markedly just before emulsion breakage occurs.   The proper setting is slightly wider than the opening which results in this sudden increase in viscosity.   The correct opening is usually found within the range of 25 to 40 mil.

It should be possible to set up the Dixie-Charlotte system on a continuous rather than semicontinuous basis.   This would call for adding the egg component suspended in the oil and the sugar and salt dissolved in properly diluted vinegar continuously in correct proportions to the partially filled Dixie mixer.   The mix would be fed continuously to the Char-

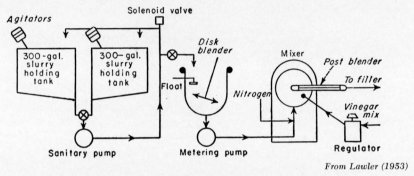

From Lawler (1953)

FIG. 22.   AMF MAYONNAISE SYSTEM USING OAKES MIXER

Reprinted from *Food Engineering*

lotte mill at the same rate of flow.   The egg-oil suspension might require premilling to obtain a workable emulsion.   Any spice or spice oil ingredient could be incorporated with the egg component.

AMF System.—The AMF system is shown in schematic form in Fig. 22.   The equipment consists of a premix feed tank and two mixing stages (Lawler 1953; Joffe 1956).   The premix supply is maintained under agitation to prevent separation of the components, but mixing is gentle in order not to thicken the suspension of egg in oil.   The first emulsification stage is designed for continuous operation, unlike the Dixie mixer.   The AMF first stage consists of a small bowl equipped with a turbine agitator rotating at 875 rpm.   The premix, a suspension of egg, oil, dry ingredients and up to 20% of the dilute vinegar solution, is whipped into a thick but coarse emulsion in the first mixer.   It is then pumped to the second stage, an Oakes type mixer.   The balance of the vinegar solution is injected into the emulsion stream as close as possible to the Oakes head or directly into it.

The Oakes mixing head normally rotates at about 475 rpm. This is slow in comparison to the Charlotte mill, but the mixing action of the Oakes head with its intermeshing teeth is very intense. The oil droplets are much smaller and more uniform in size than are obtained from the Dixie-Charlotte system. The finer oil dispersion gives mayonnaise made by the AMF system a brighter appearance than product made by other methods.

Miscellaneous Mixing Devices.—Other equipment can produce a mayonnaise emulsion (Finberg 1955). Such equipment may be used to re-

From Bolanowski (1967)

FIG. 23.   VOTATOR CR MIXER

Mixer is opened to show front stator and rotor. The pumping section is immediately behind the front rotor face. A second rotor face and opposing stator are directly behind the pump.

Reprinted from *Food Engineering*

place the premixer, the colloid mill, or both. It seems, however, that a two-stage mixing system produces the smoothest and most stable emulsions.

Shear pumps are used for manufacture of mayonnaise. The Girdler CR Mixer (Fig. 23) is this type of machine. It is a combination of an Oakes mixer and an internal pump to circulate the product. Figure 24 shows a schematic diagram of a mayonnaise plant using the CR mixer (Bolanowski 1967).

Ultrasonic vibrations are claimed to emulsify mayonnaise. The Sonolator (Fig. 25) has been used as a means of producing such vibrations. It is operated by forcing a stream of fluid against the edge of a knife-like blade which vibrates on impact of the fluid (McCarthy 1965). Figure 26

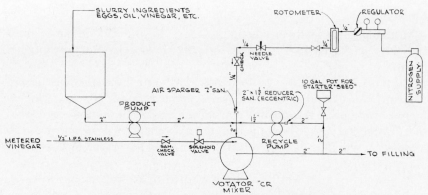

Courtesy of Votator Div., Chemetron Corp.

FIG. 24. MAYONNAISE AND SALAD DRESSING SYSTEM USING VOTATOR CR MIXER

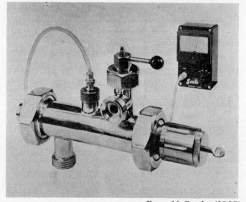

From McCarthy (1965)

FIG. 25. ULTRASONIC VIBRATION GENERATOR

Blade adjustment screw is on right, pressure
regulating valve, top center, and acoustic intensity meter in the background.
Copyright © (1965) by *Official Digest, Federation of Societies for Paint Technology*

is a diagram of the Sonolator, illustrating placement of the jet orifice opposite the vibrating element. Fluid pressure, shape of orifice, and distance between orifice and reed are critical. They will vary with viscosity
of the fluid to be emulsified.

A number of mills are available which perform as colloid mills. Some
are called such, others are known by trade name. These would be used
in the same manner as the Charlotte mill. They may be more or less effective depending on their design.

Packaging

Mayonnaise is usually packed in glass in 8, 16, and 32 fluid ounce sizes. It is also packaged in 1-gal. glass or polyethylene jars for institutional trade.

Mayonnaise is sensitive to oxidation. After being sealed in the jar, mayonnaise will oxidize until the headspace oxygen is used up. The extent of oxidation at this point is not sufficient to result in off-flavor formation. After the jar has been opened, the mayonnaise will be reexposed to oxygen. It will then develop reverted and rancid flavors unless it is stored in the refrigerator.

A loose lid or a lid which has been warped by overtightening will leak and admit oxygen from the atmosphere causing rancidity. Oxidation or

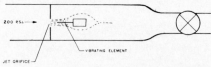

From McCarthy (1965)

Fig. 26. Schematic Diagram of Ultra-sonic Vibration Generator

Copyright © (1965) by *Official Digest, Federation of Societies for Paint Technology*

moisture loss from the emulsion through such leakage will also cause the emulsion to darken on the surface and eventually break.

Polyethylene jars are permeable to oxygen. Mayonnaise packed in such jars will oxidize and become rancid in about two months. Use of polyethylene is restricted to institutional product where a high turnover rate is assured. Mayonnaise for grocery distribution may be on the shelf for too long a time to permit the safe use of polyethylene containers.

Overfilling the jar will cause the lid to force mayonnaise emulsion over the closure threads. The emulsion is exposed to the atmosphere and will soon dry out. The oil residue will oxidize to the point of forming a gel with a strongly rancid odor and flavor. The gummy residue will be disagreeable to the touch on opening the jar. It will also affect the odor and flavor of the jar contents, at least on the surface. Obviously, overfill is to be avoided.

Filling equipment is adjustable as to fill volume. A thick mayonnaise emulsion which has been fluffed with nitrogen or carbon dioxide may hold the gas under pressure until after it has been filled into the jar. It will then expand slowly, flooding the jar before it can be capped. This type of overfill can be corrected by adjusting the filling machine to give an apparent underfill at that point.

Quality Measurements

Mayonnaise quality is difficult to assess properly. Some physical measurements have been applied but with limited success.

Kilgore (1930) describes a form of penetrometer known as the "Plumit." This consists of an aluminum cone with a graduated rod protruding from the base of the cone. The Plumit weighs 14.5 gm and is 13 cm in length. In use, the Plumit is held vertically by the top of the rod so that the cone point is aimed at the surface of the mayonnaise in the jar. The top of the rod is held exactly 12 in. above the mayonnaise surface. The Plumit is then dropped into the mayonnaise. The depth to which it penetrates is read from the graduated stem to the nearest 0.5 cm. Penetration depth correlates inversely with viscosity.

Viscosity can be measured directly with a Brookfield Helipath Viscometer (Anon. 1965). An ordinary viscosimeter will not give meaningful readings as movement of the normal viscosimeter measuring device would deform the emulsion irreversibly. The Helipath instrument consists of a short horizontal wire welded to a vertical wire. The vertical wire is rotated by the Brookfield Viscometer while the entire Viscometer is being lowered slowly into the jar of mayonnaise. The horizontal wire traces a spiral or helical path through the sample so that it is constantly being exposed to fresh, undisturbed product. Resistance by the product to movement of the wire is recorded as a number which can be correlated with viscosity in centipoise units if so desired.

Plumit or Helipath viscosity can describe variations in any given mayonnaise. It can be used to accept or reject product on a fairly sound basis as long as the formula and production methods are not changed markedly. Viscosity becomes meaningless in attempting to relate mayonnaises of different formulation and process history to each other. Other attributes enter into the picture.

One such attribute may be called "body." Two mayonnaise products can have identical viscosity measurements. However, one can be removed from its container a spoonful at a time without change in firmness throughout the operation. The other mayonnaise could become pourable before half the jar is emptied. The first has a firm body, the second, a soft one.

Texture is another attribute. A jar of mayonnaise is usually dry in appearance within 24 hr after it is first made. The mayonnaise should also be a rigid gel. On storage, the dryness and rigidity will show slow transformation to a softer and glossier texture. Such change will become apparent within 1 or 2 months.   However, the Plumit or Helipath viscosities can remain constant throughout these obvious textural changes.

Apparent flavor is also related to body and texture but not to viscosity. Two mayonnaise emulsions with identical salt, vinegar, and sugar levels and the same viscosity could be entirely different to taste, one being bland, the other sharp.

No mechanical measurements have been found to replace the tactile evaluations of texture, body, and taste.   Experience and a tablespoon are apparently still required.   Arbitrary numerical scales can be used to record observations.   As with all sensory evaluations, a panel of several observers should be used in place of a single person.

Real or simulated shipping tests are used for mayonnaise evaluation. Cases of product to be evaluated are shipped in normal fashion to some distant point and returned.   A round trip of 1,000 miles is practical.   Some devices are available which bounce the test product with a definite force at a regular interval to simulate long distance shipping.   One recommendation is for a 1.5-in. drop at the rate of 1050 per hr (Anon. 1965). Changes in viscosity, body, texture, and headspace in the jar as a result of this mechanical shock are noted.   If they are excessive, the product is judged to have a weak emulsion and the causes are sought out to be corrected.

SPOON TYPE SALAD DRESSING

Spoonable salad dressing was first developed as a low cost substitute for mayonnaise.   It was to resemble a cross between boiled dressing and mayonnaise (Anon. 1933).   Spoonable dressings are now accepted as a different product.   The market is divided into areas where mayonnaise is preferred to salad dressing, e.g., the Southeast, and other areas such as the Midwest where salad dressing is the preferred product.   Salad dressing often commands as high a price as mayonnaise.   Salad dressing is tart and tangy.   Mayonnaise is relatively bland and more subtle in flavor. Mayonnaise and salad dressings are used in similar fashion on sandwiches, and in potato, tuna, and chicken salads, etc.   The choice between the two is one of personal taste and not of quality or relative cost.

Standards of Identity

Standard salad dressing is described in terms similar to those used for mayonnaise.   It too is an emulsified semisolid food.   This type of salad dressing is referred to as spoonable to distinguish it from the pourable dressings.   These latter products, with the exception of French dressing, are not covered by Standards of Identity.

Salad dressing under the Standards resembles mayonnaise in that it is an emulsion of oil in vinegar using egg as an emulsifier. It differs from mayonnaise in that it also contains starch paste as a thickener.

The oil, egg, vinegar, and the optional ingredients, salt, sweeteners, spices, flavorings, EDTA salts, and monosodium glutamate are described by the salad dressing standards in identical language to that used for the same ingredients in mayonnaise. There is, however, no limitation on the level of total acid in vinegar or the citrus juices.

Salad dressing may also contain not more than 0.75% by weight of any of several polysaccharide gums or the cellulose derivatives, methylcellulose and sodium carboxymethyl cellulose. The gums may contain 0.5% dioctyl sodium sulfosuccinate as a solubilizing agent. Gums are not permitted in mayonnaise.

TABLE 22

SPOONABLE SALAD DRESSING COMPOSITION

Ingredient	Weight %
Salad oil	35.0–50.0
Fluid egg yolk[1]	4.0– 4.5
Salt	1.5– 2.5
Sugar	9.0–12.5
Mustard flour[2]	0.2– 0.8
Starch	5.0– 6.5
Vinegar (100 gr)	9.0–14.0
Spices[2]	
Water to make 100%	

[1] Egg solids 43%. May substitute whole or fortified egg, fluid or dry, on a total solids basis.
[2] Spice oils, oleoresins may be substituted.

The Standards require not less than 30% vegetable oil nor less than 4% liquid egg yolk by weight. When whole or fortified egg or dried egg is used, the yolk solids content must be equivalent to that obtained from 4% minimum of 43% solids yolk.

It is possible to make salad dressing with less than 30% oil and less than 4% egg yolk. Polysorbate 60 is permitted at 0.3% maximum in eggless salad dressing under the U.S. FDA Food Additives Amendment. Such dressings are nonstandard. They are usually formulated as dietetic foods, either as "low calorie" or "eggless" types.

Formulation and Ingredients

Formulation.—Most spoonable salad dressings are formulated within the limits given in Table 22.

As with mayonnaise, each of the components of salad dressing has a specific function. Many of the ingredients are common to both products.

They do not necessarily have the same function in each product, however. The major discussion of these ingredients will be found in the appropriate section under mayonnaise. Additional discussion with special reference to salad dressing will be made where needed.

Oil.—Oil is the major source of mayonnaise viscosity and body. Starch paste has this function in salad dressing. Oil modifies the mouth-feel of the starch paste, making it smoother and richer. Viscosity is altered by the oil component but it is affected more by the type and content of starch than by the oil level.

Starch.—The Standards permit any food starch and tapioca, wheat, or rye flour. Modern salad dressings are usually made with blends of corn and modified waxy maize starch. Potato and tapioca starches may be used for special dressings. Other starches, such as arrowroot, have become too costly to be competitive.

Starch is a polysaccharide consisting of glucose molecules linked in a chain. The starch chains may be straight, in which case the starch component is called "amylose," or they may be branched and called "amylopectin." Amylose and amylopectin behave differently from each other when made into starch paste.

Starches may be further modified by treatment with phosphate salts and by reaction with various chemicals which crosslink the normal polysaccharide chains (Hamilton and Paschall 1967; Hullinger 1967). A wide variety of characteristics can be obtained by blending different natural and modified starches.

Corn starch is composed of 27% amylose and 73% amylopectin (Watson 1967). Paste made from corn starch has a low viscosity when hot, but sets to a firm gel on cooling. If the gel is disturbed after it has set, the structure of the gel is permanently altered and may break down completely. Corn starch is sensitive to high acidity. Salad dressing made with such starch will become softer with age on storage.

Waxy maize, waxy sorghum, and tapioca starches are 100% amylopectin. Starch paste from these starches have a high viscosity when hot as well as when cold. They do not gel. Amylopectin starch pastes are thixotropic in that they soften when disturbed but recover their original viscosity on being allowed to stand quietly for a few hours. These pastes do tend to be stringy. They also soften on being exposed to the acids of salad dressing but to a lesser degree than does corn starch paste (Hullinger 1967).

The crosslinking of starches makes them less susceptible to acid breakdown. It also reduces the gelling tendency of corn starch. Salad dressing starches are usually blends of crosslinked and unmodified starches in various proportions to give them special characteristics (Anon. 1966). They are sold under brand names or are specially blended for individual salad dressing producers.

In general, each starch processor has two regular blends for dressing use. One of these has a high level of modified waxy maize starch for smooth creamy texture and good resistance to mechanical shock. It is most suitable for batch cooking or for continuous cookers which can handle high hot viscosity paste. The other blend is designed for operations which have continuous cookers that are designed for handling low hot viscosity paste. This blend contains a high proportion of modified corn starch for low viscosity during cooking. It is prone to gel to some extent and will, therefore, be less resistant to mechanical shock in handling and shipping.

Phosphate modified starches have good freeze-thaw stability (Hamilton and Paschall 1967). This property was developed primarily for frozen pies and gravies. Krett and Gennuso (1963) and Partyka (1963) have developed freezable salad dressings which make use of freeze-thaw stable starches.

Egg.—The emulsifying properties of egg yolk are essential for maintaining oil in proper distribution in salad dressing. Other emulsifiers may be capable of duplicating the performance of yolk but are not permitted under the Standards. Starch has the effect of forming the emulsion by mechanical means but requires egg yolk to make the dispersion permanent.

Egg white proteins in whole or fortified egg acts as an emulsifier when it is coagulated by the acid in salad dressing (Flückiger 1966). Milling the dressing disperses the egg white, modifying the starch texture. The effect is to produce a drier, less pasty product.

Acid.—As with mayonnaise, acid is the only preservative permitted against the growth of microorganisms in salad dressing. The ability of the acid to prevent growth is dependent on its concentration in the aqueous phase.

The perception of sour taste is also dependent on the concentration of acid in water. Mayonnaise has sufficient oil to coat the tongue and mask this tartness. Salad dressing does not. Its tartness increases with decrease in oil content: The total acidity must increase as the proportion of water increases.

Seasoning.—The higher water content of salad dressing requires that much more sugar be added to its formulation than is used in mayonnaise. Flavor balance involves the varied sensitivity of the palate to different concentrations of flavor components. A rule-of-thumb guide calls for sugar and 100 gr vinegar in about equal proportions. Only slight variations from this are required to change the emphasis in sweetness or tartness.

Salt does not follow the same rule. Salad dressing has only a little more salt than does mayonnaise. A slight increase in salt content over these values would make a noticeably saltier product. Careful tasting of mayonnaise will reveal that it is quite salty. An equivalent concentration of salt in salad dressing would make it unbearably salty to taste.

The choice between mustard flour and mustard oil is potentially more critical for salad dressing formulation than for mayonnaise. Starch paste structure is affected by the proteins and other mustard flour components in the same manner that it is affected by egg white and oil. The same is true of other spices and herbs but these are used in lower quantity and are, therefore, less effective.

Salad dressings are more highly seasoned than is mayonnaise. Spice blends are closely guarded secrets. Flavor houses offer a number of these and are willing to modify standard blends to order.

Spice oils are preferred by some dressing processors as many natural spices are highly colored. Other processors prefer whole spices. Small amounts may be used if finely ground. Natural spices have a fuller flavor than oils extracted from them. Care must be taken that whole spices are clean as they may be a source of excessive mold and bacterial contamination. The use of EDTA salts would also be desirable where whole spices are used. This would eliminate the prooxidant effect of trace metal contamination from the spices. These metals accumulate during the grinding operation.

The spices most frequently recommended for salad dressings are mustard, black pepper, tarragon, onion, garlic, allspice, and clove (Cumming 1964). Paprika is added for color and may be as the oleoresin or as a finely ground powder.

Gums.—The gums permitted under the Standards are described as "optional emulsifying ingredients." They are not emulsifiers in the strict sense. Emulsifiers are compounds which have water soluble groups attached by chemical bonding to an oil soluble carbon chain, often a fatty acid residue. Gums are water dispersible colloids and have no oil soluble portion. They are called protective colloids. They behave similarly to starch paste in preventing oil droplet coalescence. They offer a viscous physical barrier against emulsion breakage but require a true emulsifier to form and permanently stabilize that emulsion.

Gums must be used with discretion in salad dressing. Excessive amounts tend to develop a sliminess along with increase in viscosity. They are not used in spoonable dressings as frequently as they are in the pourable salad dressings.

Equipment and Process

Fundamentals.—The basic process for spoonable salad dressing manufacture is similar to that used for mayonnaise. The major difference is that starch paste is incorporated into the egg-oil blend in place of vinegar solution. Most formulations for salad dressing are separated into those two parts, the egg and oil blend and the starch paste.

Raw starch is not water soluble. It disperses readily but requires cooking to form a thickened suspension or paste. The sugar, salt, and vinegar are also part of the paste formulation.

The process calls for cooking and cooling the paste separately before combining it with the egg and oil emulsion. In batch processing, some of the paste is usually added to the egg before any oil is worked in. This weakens the egg. The oil is thereby prevented from forming too tight an emulsion. The oil is not whipped in as vigorously as with mayonnaise. When the balance of the starch paste is mixed in after the oil is added, the entire mass must be fairly soft and pumpable. The colloid mill gives salad dressing its final high viscosity and smooth texture. If the premix is too stiff, it will break down in the mill.

In the Dixie-Charlotte system, salad dressing seed remaining from the previous batch of premix will soften the egg sufficiently. It is not normally necessary to add more paste to the egg before proceeding. The Charlotte mill setting should be more open for salad dressing than for mayonnaise. A range of 45 to 50 mils is usual.

The Dixie system consists of jacketed kettles with sweep agitators added to the mayonnaise equipment (Fig. 27). Starch paste is cooked in 1 and cooled in 2 or 3 other kettles alternately. The cooled starch paste is pumped into the Dixie premixer. Starch paste may be cooked and cooled in the same kettle where the operation is sufficiently small to do without continuous processing.

In other continuous systems, it is feasible to pump egg, oil, and starch paste into a mixing zone in separate streams. These are preblended in the mixing zone to be pumped to a colloid mill for finishing (Lipschultz and Holtgrieve 1968).

No matter what process is used, starch paste should be used as soon as possible after preparation. The longer it stands after cooling, the more readily it will break down on being worked into the dressing. This would be especially true with high amylose starches.

Starch Paste Preparation.—Starch begins to cook between 185° and 195°F depending on the type of starch involved. It remains practically unchanged below its cooking temperature. Viscosity of the paste increases to a maximum as cooking continues. This requires 4–6 min at

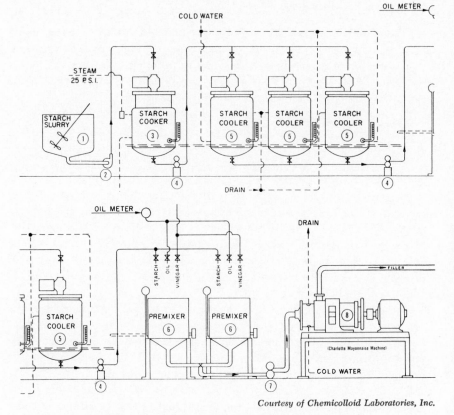

Courtesy of Chemicolloid Laboratories, Inc.

FIG. 27.   DIXIE-CHARLOTTE SALAD DRESSING SYSTEM

Upper panel of schematic is for preparation of starch paste. This is added to the
mayonnaise unit depicted in the lower panel.

atmospheric pressure and somewhat less time under higher pressure.
The paste is then cooled both to stop the cooking action and to prepare it
for further use.   Figure 28 gives the viscosity curves for two different
types of starch during the cooking and cooling process.   These are a low
hot viscosity modified starch with high amylose content and a modified
amylopectin (waxy maize) with high hot and cold viscosities.   Low
viscosity starch is better suited to plate heat exchanger cooking systems
where high pressures are undesirable.

Cooking of starch in vinegar solution tends to degrade or hydrolyze
the starch to some extent.   Modified starches are usually more resistant
to breakdown than the unmodified types.   Where feasible, it is best to
cook the starch with a minimum amount of vinegar in the water solution.

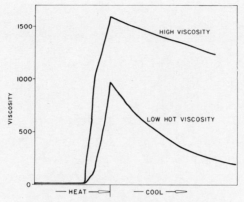

Courtesy of National Starch and Chemical Corp.

FIG. 28. VISCOSITY OF STARCH PASTE

Viscosity curves for two typical salad dressing starches determined by the Brabender Visco-Amylograph. This instrument has a time cycle which measures viscosity during both cooking and cooling of starch paste.

The balance of cold vinegar can then be added on completion of cooking to help cool the paste. The limiting factor on reducing vinegar during cooking is the maximum viscosity which the cooking equipment can handle. This modification in procedure is usually limited to batch cooking.

Continuous cooking equipment must allow for holdup time during the process so the hot starch can cook fully before it is cooled. Pressure can reduce this time to some degree. Excessive pressure or excessive time under pressure can degrade starch, especially the unmodified types.

Pregelatinized starch prepared by cooking a paste and drying it has been used for salad dressing preparation. It must be properly rehydrated for use. This requires holding the reconstituted paste for at least 1 hr before adding it to the oil suspension. Pregelatinized starch is too costly to use on a regular basis. It is good for emergencies caused by cooking equipment failure or by temporary production requirements in excess of available starch cooking capacity.

Starch Cookers.—Batch cooking is the simplest procedure for preparing starch paste. Practically any jacketed, agitator-equipped covered kettle can serve as a cooker. Sweep agitators are preferred. The Dixie-Charlotte starch cooker is shown in Fig. 29. The jacket should be fitted with steam and cold water lines for heating and cooling. The cover is used to minimize water and vinegar loss from vaporization during cooking.

Courtesy of Chemicolloid Laboratories, Inc.

FIG. 29. DIXIE STARCH COOKER
SHOWING SWEEP AGITATOR

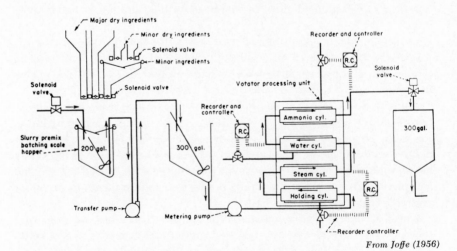

From Joffe (1956)

FIG. 30. VOTATOR STARCH COOKING SYSTEM

Output of starch paste is fed to holding tanks and from there to the premix tank.

Reprinted from *Food Engineering*

The Votator heat exchanger is one of the most commonly used continuous cookers for starch paste. The system consists of heating, holding, and cooling units connected in a series (Fig. 30). Holdup time for complete cooking is about 15 sec (Ziemba 1949). The Votator can handle high hit viscosity starch pastes.

An automatically controlled steam injection cooker has been developed

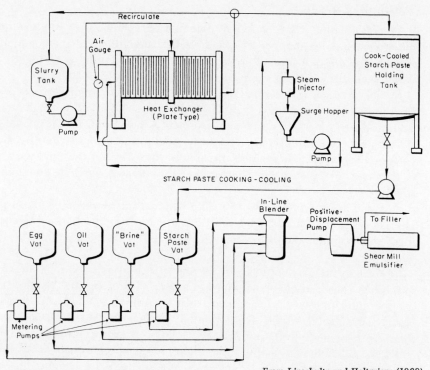

From Lipschultz and Holtgrieve (1968)

FIG. 31. CONTINUOUS MAYONNAISE AND SALAD DRESSING SYSTEM

This system shows use of plate heat exchanger for cooking and cooling of starch paste.

Reprinted from *Food Engineering*

(Anon. 1964). Holdup time is determined by the length of pipe inserted between the steam injector and a cooling device.

Plate heat exchangers are used for cooking starch paste. Lipschultz *et al.* (1968) describe a plate heat exchanger system combined with steam injection. Cooling of the paste is accomplished in a second section of the plate heat exchanger (Fig. 31). Heavy plates are required in order to withstand the high pressures which develop on forcing viscous paste through the equipment. Ordinary plates would bend under such pressure and cause leakage of product.

Packaging and handling of salad dressing is the same as for mayonnaise. Evaluation of quality is also similar.

FRENCH DRESSING

Standards of Identity

French dressing is the basic pourable salad dressing. It is a mixture of oil, vinegar, and a number of spices. Other ingredients may be added. French dressing is manufactured under U.S. FDA Standards of Identity. Other pourable salad dressings are not covered by such standards.

French dressing is described as a separable liquid or emulsified viscous food prepared from not less than 35% edible vegetable oil by weight and acetic or citric acid. The oil may contain a maximum of 0.125% oxy-stearin as a crystal inhibitor. The acetic acid in the form of vinegar may contain citric acid in an amount not greater than 25% of the total weight of the acids. Lemon or lime juice may be used as an alternate source of acid.

Optional flavoring ingredients include salt, natural sweeteners, mustard, paprika and other spices, spice oils and other naturally obtained flavorings, monosodium glutamate, tomato paste, purée or catsup, and sherry wine.

Emulsifiers are also optional since French dressing may be separable or emulsified. The latter type is more common. The permitted emulsifiers are the gums arabic (acacia), locust bean (carob), guar, karaya, tragacanth, carrageenan (Irish moss), pectin, propylene glycol alginate, sodium carboxymethylcellulose, methyl cellulose, hydroxypropyl methylcellulose, calcium carbonate, and sodium hexametaphosphate, and any combination of them. The gums may contain 0.5% by weight dioctyl sodium sulfosuccinate as a solubilizing agent. Liquid, frozen, or dried egg yolk, fortified yolk, or whole egg may be used. The total emulsifier level is restricted to a maximum of 0.75% by weight of finished dressing. The egg ingredient is limited to the amount containing 0.75% yolk solids since these solids constitute the sole emulsifying ingredient in eggs.

Calcium disodium or disodium EDTA is permitted at 75 ppm maximum as a metal scavenger. The inert gases, nitrogen and carbon dioxide, are permitted for packaging.

Formulation and Process

The quality of French dressing is partially dependent on the amount of oil used in the formula. Although a minimum oil level of 35% is permitted, 55–65% is the usual. This results in a satisfying mouth-feel, especially in emulsified types. Less oil emphasizes the gum component and may contribute a slimy character to the dressing.

French dressing is popularly conceived of as being red in color. This is not required but is most frequently the case. Paprika is the major source of color. Some dressings contain tomato solids as well but most do not. EDTA salts are used to protect this color from fading.

Sugar, salt, and vinegar levels are a matter of taste. Sufficient vinegar must be used for preservation. Each processor has his own interpretation of what the market will prefer. Garlic is one of the major flavorings. The amount used is also a matter of taste and judgment of the manufacturer.

Although a number of polysaccharide gums are permitted, propylene glycol alginate is used most frequently. Gum tragacanth is next in popularity. These gums are probably less affected by acid than are the other gums and therefore contribute a longer shelf-life to the dressing. When gums are used, it is essential that they be properly hydrated before the product is milled. Poorly hydrated gums will not stabilize the emulsion effectively.

The various components of separating type dressings must merely be well-mixed before bottling. If the product is to be emulsified, it is passed through a colloid mill. The Dixie-Charlotte system is frequently used.

Packaging

The "banjo" bottle has become traditional for French and other pourable dressings. The shape is modified slightly from brand to brand to achieve some individuality. The bottle is sealed with a metal cap.

Polyethylene cannot be used as it is permeable to oxygen. Dressings would become rancid rapidly in a polyethylene container. One brand of emulsified French dressing was found with a hollow polyethylene plug as a closure for a glass decanter bottle. The dressing in the neck of the bottle was bleached in each sample observed due to the penetration of oxygen through the plastic plug.

NONSTANDARD DRESSINGS

Pourable Dressings

A large variety of emulsified and separating pourable dressings are available. They resemble French dressing in many respects but depart from the standards in any of several ways. Some contain Polysorbate 60 emulsifier. If the vinegar level is reduced to decrease tartness, sodium benzoate or potassium sorbate is added as a preservative. Artificial flavors are often used. They also contain chopped onion, pepper flakes, and other vegetables. Some of these dressings are sold as "Italian," "Russian," Blue Cheese, or Roquefort. Others have exotic names, partially to describe a flavor concept, partly to attract attention.

A number of low calorie emulsified pourable dressings were observed in a retail grocery. They invariably contained less oil than is used in regular dressings. Viscosity was achieved by increasing the gum content. Ultrasonic vibrations seem to help disperse oil in such dressings (Samuel 1966). The satisfying mouth feel of regular dressings is sacrificed for reduction in caloric content. Table 23 gives the nutrient composition of some of these dietetic dressings as taken from their label declarations.

TABLE 23

LOW CALORIE SALAD DRESSING COMPOSITION

	Weight %				
	Italian		Imitation French		
Ingredient	1	2	1	2	Blue Cheese
Oil	2.50	5.7	6.5	10.74	7.2
Protein	0.08	0.3	0.6	0.47	5.1
Carbohydrate	6.92	3.6	20.0	14.10	2.5
Calories per tsp	3	3.5	8	7.5	5.5

TABLE 24

LOW OIL SPOONABLE SALAD DRESSING

Ingred ent	Weight %
Water	69.22
Oil	10.00
Vinegar (50 gr)	9.60
Avicel (microcrystalline cellulose)	5.00
Cooked starch	2.00
Salt	1.80
Glycerol monostearate	1.50
Gum tragacanth	0.30
Mustard flour	0.15
Polysorbate 60	0.15
Saccharin	0.005
Sodium benzoate	0.05
Potassium sorbate	0.05
Onion powder	0.03

Source: American Viscose Div., FMC Corp.

Spoonable Dressings

A low calorie spoon type dressing formula was developed using edible cellulose powder as a nonnutritive thickener (Anon. 1968). The dressing formula is given in Table 24. It is processed in the same manner as standard spoonable salad dressing. Milling is required to smooth the cellulose

TABLE 25

HEAT STABLE SALAD DRESSING

Ingredient	Weight %
Water	36.72
Oil	30.00
Vinegar (50 gr)	14.00
Sorbitol solution	10.00
Avicel (microcrystalline cellulose)	5.50
Salt	1.50
Starch	1.00
MSG	0.20
Polysorbate 60	0.30
Mustard flour	0.30
Gum tragacanth	0.30
Onion powder	0.15
Garlic powder	0.02
White pepper	0.01

Source: American Viscose Div., FMC Corp.

powder texture as well as to disperse the oil. It should be noted that the dressing contains no egg and therefore may contain Polysorbate 60 as an emulsifier.

Heat Stable Dressing

A sterilizible salad dressing has been developed using microcrystalline cellulose as a stabilizer. The dressing is said to withstand 270°F for 75 min without breaking or separating. The cellulose will not darken or lose viscosity on being heated under those conditions. Canned salads prepared with this dressing have been marketed. The formula is given in Table 25.

BIBLIOGRAPHY

ANON. 1933. Miracle Whip. Food Ind. 5, 324.

ANON. 1964. System increases starch cooking efficiency. Food Eng. 36, No. 4, 110.

ANON. 1965. Tips on making better dressings. Food Eng. 37, No. 2, 88–89.

ANON. 1966. Explains how starches affect dressing quality. Food Eng. 38, No. 2, 133-134.

ANON. 1968. Low fat salad dressing. Food Eng. 40, No. 8, 49.

BICE, C. 1965. Spices and mustard in today's dressings. Mayonnaise and Salad Dressing Inst., Chicago, Ill.

BOLANOWSKI, J. P. 1967. Mixing that's "tuned in." Food Eng. 39, No. 10, 90–93.

BROWN, L. C. 1949. Emulsion food products. J. Am. Oil Chemists' Soc. 26, 632–636.

CUMMING, D. 1964. The spicing of nonpourable dressings. Food Technol. 18, 1901–1902.

FINBERG, A. J. 1955. Advanced techniques for making mayonnaise and salad dressing. Food Eng. 27, No. 2, 83–91.

FLÜCKIGER, W. 1966. On the technology of mayonnaise and mayonnaise-like emulsions. Fette, Seifen, Anstrichmittel 68, 139–145. (German)

HAMILTON, R. M., and PASCHALL, E. F. 1967. Production and uses of starch phosphates. In Starch: Chemistry and Technology, R. L. Whistler, and E. F. Paschall (Editors). Academic Press, New York.

HULLINGER, C. H. 1967. Production and use of cross-linked starch. In Starch: Chemistry and Technology, R. L. Whistler, and E. F. Paschall (Editors). Academic Press, New York.

JOFFE, M. H. 1942. Mayonnaise and salad dressing products. The Emulsol Corp., Chicago.

JOFFE, M. H. 1956. How automation techniques can robotize mayonnaise, dressings. Food Eng. 28, No. 5, 62–65.

KILGORE, L. B. 1930. Introducing the "plumit." Glass Packer 4, 65–67, 90.

KILGORE, L. B. 1932. The mustard and the mayonnaise, flavor ingredient has important effect on stability of the emulsion. Glass Packer 11, 621–623.

KILGORE, L. B. 1933. Effects of mustard on permanency, stability and consistency of mayonnaise. Glass Packer 12, 97–99.

KILGORE, L. B. 1934. Testing mustard for mayonnaise. Glass Packer 13, 114–116.

KILGORE, L. B. 1935. Egg yolk "makes" mayonnaise. Food Ind. 7, 229–230.

KRETT, O. J., and GENNUSO, S. L. 1963. Salad dressing. U.S. Pat. 3,093,486. June 11.

LAWLER, F. K. 1953. Quality plus efficiency. Food Eng. 25, No. 6, 44–46.

LIPSCHULTZ, M., and HOLTGRIEVE, R. E. 1968. Multi-process makes 25 dressings. Food Eng. 40, No. 11, 86–87.

LIPSCHULTZ, M., PISTORIUS, K., and JENSTEAD, R. L. 1968. Precision ensures quality. Food Eng. 40, No. 4, 87–89.

MATTIL, K. F. 1964. Cooking and salad oils and salad dressings. In Bailey's Industrial Oil and Fat Products, 3rd Edition, D. Swern (Editor). Interscience Publishers Div., John Wiley & Sons, New York.

McCARTHY, W. W. 1965. Pigment dispersions by sonic techniques. Offic. Dig. Federation Soc. Paint Technol. 37, 1650–1660.

PARTYKA, A. 1963. Salad dressing. U.S. Pat. 3,093,485. June 11.

REYNOLDS, M. C., and HARRIS, B. R. 1932. Frozen egg products. U.S. Pat. 1,842,733. Jan. 26.

SAMUEL, O. C. 1966. Tunable ultrasonic homogenizer boosts product marketability. Food Process.-Marketing 27, No. 7, 82–84.

SLOSBERG, H. M. 1968. Personal communication. New York, N.Y.

WATSON, S. A. 1967. Manufacture of corn and milo starch. In Starch: Chemistry and Technology, R. L. Whistler, and E. F. Paschall (Editors). Academic Press, New York.

ZIEMBA, J. V. 1949. High efficiency attained on multi-product line. Food Ind. 21, 292–294, 423–424.

CHAPTER 11

Peanut Butter

Peanut butter originally consisted of roasted, ground peanuts with a small amount of salt added for flavor. It was prepared commercially by the individual shopkeeper. Nuts were roasted daily and ground to order or shortly ahead of time in anticipation of sales. The product was meant to be consumed within a few days after preparation. It was packaged in folded paperboard cartons or boats in much the same way as bulk potato salad is still sold from the delicatessen counter.

This type of peanut butter is sold today under the label "Old Fashioned." Many changes have been made in peanut butter formulation and production since its inception. Some changes were adopted to make feasible large-scale production and widespread distribution. This meant the development of various stabilizers to prevent oil separation.

Sweeteners were added to peanut butter. Old-timers among the peanut butter processors and purists consider this an adulteration. They argue that sweeteners are a cheap filler and substitute for expensive peanuts. They overlook the preference for a sweeter peanut butter by its major consumer, children.

STANDARDS OF IDENTITY

After much deliberation and struggle with the peanut butter industry, U.S. FDA Standards of Identity were promulgated for peanut butter in 1968. The Standards are stated in simple form. Peanut butter is defined as a food prepared by grinding shelled, roasted peanuts and adding not over 10% other ingredients on an optional basis. The optional ingredients must perform a useful function.

The optional ingredients normally include salt, natural sweeteners, emulsifiers such as lecithin and hydroxylated lecithin, and stabilizers, e.g., fully hydrogenated vegetable oils and hard monoglycerides. Peanut oil may be removed or added as either liquid or partially hydrogenated oil. Such oil, when added, may be considered as part of the peanut component but the total oil content of the butter must not exceed 55% by weight.

Addition of vitamins A, B, C, and D is not permitted. Artificial color and flavor, nonnutritive sweeteners, and preservatives are also not permitted.

INGREDIENTS

Peanuts

There are three major varieties of peanuts for the peanut butter processor: Runners, Spanish, and Virginias. Peanut growers, shellers, and producers of whole nut products recognize many subvarieties. Once ground into peanut butter, previous identity of the nut becomes lost.

Only U.S. Grade No. 1 peanuts and splits are used for peanut butter manufacture. The nuts are shelled, graded, and bagged by the supplier. The peanut butter producer purchases the nut on market price without regard to variety. Runners are normally the cheapest and therefore most commonly used. An occasional surplus of Spanish or Virginias brings their price down. They are then purchased for peanut butter production.

Runners and Spanish peanuts can be processed into peanut butter separately or in blends. Virginias are too low in oil content to be used alone. Oil can be added, preferably before grinding the nuts. Most frequently, however, the other varieties are blended with Virginia peanuts to offset their lack of oil.

U.S. No. 1 shelled peanuts are whole nuts in the skin. They may contain not over 3% split nuts. The splits are nuts inadvertently broken in half during shelling. The bulk of the splits are mechanically sorted from the whole nuts and bagged separately. They are sold at a discount to peanut butter producers and other processors who do not require whole nuts. There is a limitation on the proportion of splits which can be used in peanut butter, depending on the process used. This will be discussed later under Processing.

Large peanut processors purchase their entire nut supply on an annual basis. They often hold nuts in storage for 1 to 2 yr. Shelled peanuts will keep without spoilage if held at 34°–42°F and 55–75% RH (Jenkins 1968).

Sweeteners

There seems to be no agreement among various peanut butter manufacturers on a preference for sweeteners. Scanning the labels of different brands shows a wide range of carbohydrate products in use.

Corn syrup solids have the lowest sweetening power of the carbohydrates. They are manufactured by partial hydrolysis of corn starch. The sweetness level depends on the extent of hydrolysis. It is given as dextrose equivalent (DE) since complete hydrolysis yields 100% dextrose. Some peanut butter processors use the lowest available, i.e., 28 DE syrup solids for low sweetness. Others will use 38 or 48 DE solids.

Dextrose is preferred by two major peanut butter manufacturers. It has a fairly high sweetness level but not so high as to be overbearing when used at the maximum amount permitted under the Standards. Dextrose is most frequently used as the hydrated sugar, containing about 10% water by weight. Anhydrous dextrose is also available.

Sucrose has about three times the sweetening power of dextrose. It is used in conjunction with honey by one manufacturer of a nationally distributed peanut butter. The sugar in honey is invert sugar, a mixture of dextrose and fructose. This sugar is sweeter than sucrose. The moisture in honey is probably more effective in contributing a characteristic flavor to peanut butter than are the other components in the honey.

Stabilizers and Emulsifiers

Grinding of peanuts releases oil from the cells. The peanut solids will settle on standing, forming a layer of oil on the surface of the butter. A hard layer of solids will settle to the bottom of the container. It can be removed only with great difficulty once it has formed.

Stabilizers and emulsifiers are used to prevent oil separation. Stabilizers are hard fatty compounds which behave much like plasticizers in shortenings. They form a spongy matrix of fat crystals which holds the peanut solids in suspension.

Emulsifiers such as lecithin probably complex with the protein and cause it to form a more readily suspendible solid in the oil base. Peanut butter does not have sufficient moisture for emulsifiers to behave in the classic fashion. The moisture content of peanut butter is usually 0.5–2.0% (Woodroof 1966). At this level, the protein is more hydrophilic than the emulsifier.

Fully hydrogenated peanut oil was one of the first stabilizers developed for peanut butter (Mitchell 1950). At one time the peanut butter processor felt that all of the oil in peanut butter, including the fully hydrogenated oil stabilizer, should come from peanut sources. One nationally distributed butter is stabilized by pressing about 18% of the oil from roasted peanuts, partially hydrogenating the oil to a melting point of about 98°F and adding it back to the roasted peanuts for further processing (Woodroof 1966).

Fully hydrogenating oil so changes it that the oil source is no longer considered critical from a nutritional standpoint. The Standards, therefore, have permitted the use of any vegetable oil hardfat as a stabilizer. Monoglycerides prepared from any fully hardened vegetable oil source are also permitted for the same function.

Crystal structure of the stabilizer is important. Beta-crystalline hardfat such as that made from peanut oil solidifies in an unstable form when first chilled in peanut butter. As the fat solids transform to coarser, more stable crystals, the butter surface becomes dull. The butter will become less stable and might separate free oil. This would be especially true in the case of any mechanical or thermal shock to the butter. Attempts to overcome this by adding more stabilizer will result in a firmer, drier product which is undesirable.

Beta-prime hardfats and monoglycerides harden into a permanently fine grained crystal in the same manner as they do in shortenings. This gives peanut butter a glossy surface and a stability under a broad range of storage conditions. Cottonseed, soybean, and rapeseed oil hardfats, distilled monoglycerides prepared from them, and blends thereof are all recommended for stabilization of peanut butter (Sanders 1964; Japiske 1966).

The choice of which stabilizer to use is dependent on the characteristics of the particular peanut butter production system in which it is to be used. Chilling and filling equipment and conditions under which they are operated are critical factors. Trial and error methods are best for selection of the stabilizer. This should include determination of the optimum concentration for any specific stabilizer in the particular peanut butter formulation. If the optimum amount of each stabilizer under test is not used for the comparison, results of the evaluation will be misleading.

Hardfats, hard monoglycerides, and blends of these are usually available in the form of flakes or spray crystallized beads. They can be proportioned into continuous systems in that form. Some stabilizer compounds are made up of peanut oil hardfats suspended in liquid peanut oil. One of these consists of equal amounts of hardfat and oil, another of equal amounts of hardfat, salt, and oil. These compounds are pumpable fluids designed for proportioning into the peanut butter processing line (Holman and Quimby 1950).

PRODUCTS

Old-fashioned Peanut Butter

Old-fashioned peanut butter is produced on a local basis by small nut roasters in many cities throughout the United States. A few of them have a wider distribution, rarely more than statewide. Old-fashioned butter is merely roasted peanuts ground with 1.5 to 2% salt. It contains no added stabilizer or sweetener.

Free oil separates readily from unstabilized peanut butter. On separation, the oil is unprotected against oxidation. It will become rancid in a few days. Old-fashioned peanut butter should be consumed before this time. Unfortunately, much of this type of butter does not sell fast enough to remain fresh.

The amount of oil released during grinding of peanuts depends on how finely the nuts are ground. A coarsely ground butter would have less free oil to separate. Some old-fashioned butters are partially "stabilized" in this way.

Regular Peanut Butter

Most peanut butter is stabilized and sweetened. A typical formula would consist of 90% roasted peanuts, 1.6–2% stabilizer, 1.5–2% salt, and 6–6.9% dextrose or corn syrup solids.

Regular peanut butter is packed in both smooth and crunchy style. Crunchy peanut butter is made by suspending 10–15% chopped or crushed peanuts to smooth peanut butter (Avera 1966). Approximately 23% of regular peanut butter sold today is crunchy style (Tiemstra 1969).

Whipped Peanut Butter

Peanut butter can be aerated or whipped. Ordinary butter will not hold much gas unless it is stiffened in some manner. One whipped product is made firmer by increasing the stabilizer level to about 6%. The product is waxy and does not release its peanut flavor readily.

A recent development calls for stiffening ordinary peanut butter by chilling it before aeration and keeping the product refrigerated (Anon. 1968).

Commercial Peanut Butter

Peanut butter is used as a filler for packaged cracker sandwiches. This butter is ground very coarsely to minimize the free oil content. It is usually stabilized to prevent oil soakage into the cracker.

Peanut butter for candy manufacture may be similarly processed. Many confectioners and large bakeries produce their own peanut butter using their own special formulations.

Regular peanut butter is also packed in bulk for bakery and confectionery usage.

Special Products

Although the present Standards do not permit the addition of flavors to peanut butter, various flavored butters have been marketed as novelty items. Peanut butter has an intense natural flavor. As a rule, the amount of artificial flavor required to be apparent in the presence of peanut flavor is sufficient to result in bitterness and artificiality.

Unsalted peanut butter is available for those persons requiring such a product.

Peanut butter is often eaten in sandwiches in combination with fruit jelly. Much research effort was expended by many peanut butter processors in an effort to market such a combination. One manufacturer has finally succeeded in producing peanut butter swirled with jelly. It was not an easy task. Peanut butter must be practically moisture-free. Jelly has a high moisture content. Transfer of moisture from jelly to peanut butter results in a hard, pasty peanut butter and a rubbery or crystalline jelly. The solution of the problem was to provide a moisture barrier between peanut solids and the jelly. The barrier consists of a high level of monoglyceride stabilizer added to the peanut butter.

A proposed alternate solution to combining peanut butter with jelly called for preparation of a substitute for jelly (Bahr and Krumrei 1964). The substitute was actually a cream icing such as is used on bakery cakes. This icing is a colored, flavored suspension of powdered sugar in a shortening base. It contains no moisture to interfere with the peanut butter characteristics.

Soybean Butter

Cashew and other nut butters have been marketed as novelty items. The high cost of most nuts prevents their widespread use in butter form. Soybeans are lower in cost than peanuts. However, soybeans present many problems in attempting to prepare a butter from them.

Pichel and Weiss (1967) have solved a number of these problems. Soybeans have a characteristic undesirable flavor. Roasting or frying alone does not modify this flavor. Pichel found that treating soybeans with hot water, either by soaking or by steaming, practically eliminated the off-flavor.

The soybeans were then fried in oil and reduced to a fine paste. Fried soybeans are hard and brittle, unlike fried or roasted peanuts. They also lack sufficient natural oil for ordinary grinding. It is necessary to add oil to the beans in order to process them further.

Soybean solids are somewhat fibrous. Passing them through stone or steel plate attrition mills results in a paste with a chalky mouth-feel. Soy flour is similar to cocoa powder in this respect. Chocolate refining rolls can reduce such fibrous material to a smooth paste. Pichel found that he could achieve the same results with greater throughput rate by comminuting the whole fried beans on an Urschel Micro-cut mill (Hlavacek and Freedman 1968).

Sugar, salt, and stabilizers are added to soybean butter as in the preparation of peanut butter. The major drawback in butter from soybeans is in lack of flavor. Roasted peanuts have a strong, pleasant taste. Soybeans are relatively bland after removal of the undesirable flavors. Marketing of soybean butter will depend on the development of acceptable artificial flavors for such product.

EQUIPMENT AND PROCESS

Roasting

Peanuts are dry roasted as the first step in peanut butter production. Woodroof (1966) states that 40–60 min to reach a nut temperature of 320°F in an 800°F roaster are normal roasting conditions. A batch load is given as 400 lb. Roasters consist of large revolving cylinders into which the nuts are loaded. The nuts are heated by direct exposure to a gas flame or by radiant heat, in which the flame is applied to the outside of the roaster drum. Moisture is lost during roasting. The process is completed when the proper temperature is reached. It may be higher than given above if a darker than average roast is desired. The time required depends on the size and moisture content of the nuts.

On completion of roasting, the hot peanuts are dumped into a bin. Cold air is blown through the mass of nuts to cool them and prevent further roasting or darkening of the nut.

Fresh roasted peanuts have a volatile and perishable flavor. They should be processed further as quickly as possible to obtain the highest quality butter. Holding roasted nuts for several hours before grinding will result in peanut butter flavors which might be described as "stale."

Blanching

The bulk of the peanuts used for peanut butter is shelled whole nuts in the skin. Blanching is the operation used to remove this skin. The nuts are split in two causing the skin to shatter. Skin fragments are then blown away. Nuts, being heavier, remain behind.

Roasters are usually coated internally with a layer of soot accumulated from charred nut particles and free oil which oozes from the roasting nuts. The soot rubs off onto the roasting nuts. Blanching removes soot along with the skins.

Split nuts which have no skins also become coated with soot during roasting. Splits retain this soot through the entire process. The final color of peanut butter is affected by soot. The soot content and consequent gray cast in finished butter is proportional to the percentage of splits used. Some processors use no splits. Others limit themselves to 20 or 25%. Low-cost butters may contain 50–60% splits. The only difference between these butters with respect to quality insofar as split content is concerned is in the color of the butter. Split nuts are not inferior to whole nuts in wholesomeness.

Grading and Sorting

Defective peanuts fit into two categories. Immature or undeveloped nuts are small and shriveled. Mature whole nuts which have become moldy do not split during blanching. The mold causes the nut halves to stick together and the skin to remain on the nut.

Courtesy of The Bauer Bros., Co.

Fig. 32.   Steel Plate Peanut Butter Mill

Mill has been opened to show stator and rotor.

Graders and sorters are revolving cages or shaker screens with specially sized openings. Small screens are used before blanching to remove shrivels from the large whole nuts and splits. Large screens are used after blanching to hold back unsplit nuts to be discarded.

Specks in peanut butter are caused by grinding shriveled nuts and nuts with skins. This should be held to a minimum. If the specky matter were to be completely ground, the resulting butter would have an excessively bitter taste. This material apparently does not affect the flavor unduly when the particles are large enough to appear as specks.

Grinding

After being cleaned, the peanuts are ground to a fine paste. The steel burr-mill is the most frequently used mill (Fig. 32). The stone mill (Fig.

33) and a multibladed cutting mill (Fig. 34 and 35) are also used by some processors.

Stone mills and steel plate mills operate by having a rotor revolve at high speed against a stator. The distance between the faces of the rotor and stator are adjustable. Openings of 3–5 mils are normal for grinding regular peanut butter. The nuts are forced between the grinding surfaces by an impeller mounted on the rotor shaft.

Courtesy of Morehouse-Cowles, Inc.

Fig. 33. Stone Peanut Butter Mill

The upper stone is the stator, the lower is the rotor. An impeller is mounted at the top of the rotor shaft to force peanuts between the stones. Surrounding cavities are for circulation of cooling water.

Stone mills are susceptible to breakage of the grinding stones by pebbles and other hard objects in the peanut supply. The occurrence of pebbles picked up during harvest is fairly common. Peanuts must be cleaned with a great deal of care where a stone mill is the only one used. If a two-stage grinding system is used, and only the second stage is a stone mill, ordinary cleaning of the nuts can suffice. The first stage should then be a steel plate mill as this grinder will crush the pebbles without damage to the plates.

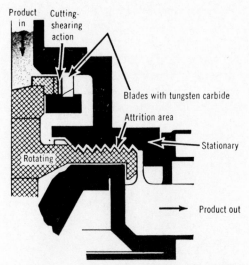

Product
in

Cutting-
shearing
action

Blades with tungsten carbide

Attrition area

Rotating

Stationary

Product out

From Hlavacek and Freedman (1968)

FIG. 34. COMBINED CUTTING AND GRINDING MILL FOR PEANUT BUTTER

Comminutor cross-section diagram shows product flow across rotor blades and vanes (crosshatched). First, cutting-shearing occurs as product passes between revolving rotor (9600 rpm) and stationary stator blades. Second, attrition occurs as product follows outward path through series of concentric vanes.

Reprinted from *Food Processing*

The multibladed cutter operates on a different principle (Hlavacek and Freedman 1968). It consists of a drum mounted on a vertical shaft rotating at 9600 rpm. The walls of the drum consist of over 200 vertically arranged tungsten carbide knives (Fig. 36). Peanuts are poured into the open top of the revolving drum. The nuts are hurled against the knives by centrifugal force and are literally slashed into a paste. Further attrition of the nut particles takes place by forcing them between a pair of steel plates. One plate is a rotor attached to the revolving drum. The other plate is a matching stator (Fig. 37). The 2-stage cutting and grinding action of this mill yields a finished butter at 150°–170°F output temperature.

Peanut butter ground through the other mills does reach 180°F if a single mill is used for all of the grinding. In actual practice, many processors use two mills in series. The first mill is adjusted for a relatively coarse grind, the second for a fine finishing grind. The output temperature of any two-stage milling operation is lower than that from a single

Courtesy of Urschel Laboratories, Inc.

FIG. 35.   ASSEMBLED CUTTING MILL FOR
PEANUT BUTTER

Courtesy of Urschel Laboratories, Inc.

FIG. 36.   ROTOR OF CUTTING MILL
SHOWING TUNGSTEN CARBIDE BLADE
ASSEMBLY

Courtesy of Urschel Laboratories, Inc.

FIG. 37. DISASSEMBLED CUTTING MILL

Steel plate rotor and stator of second attrition stage is
in center.

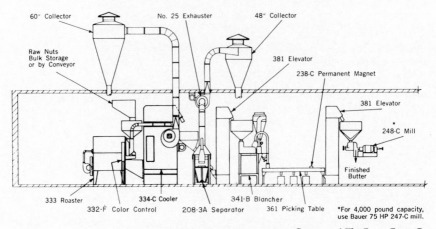

Courtesy of The Bauer Bros., Co.

FIG. 38. LAYOUT FOR INTEGRATED PEANUT BUTTER PLANT

stage mill. Normal output temperature of the second stage is 140°–170°F
depending on mill setting and throughput rate.

An integrated peanut butter plant layout is shown in Fig. 38.

There are a number of options in milling practice, each of which results in modifying the character of the finished peanut butter. These
options concern the type of sweetener added, the point in the process at
which it is added, and the temperature attained by the butter after the
sweetener is added.

Dextrose, invert sugar, and corn syrup solids are reducing sugars and
can react with free amino groups in the peanut protein. The reaction is accelerated by heat. It seems to be unusually pronounced at 180°F. Two-

stage milling tends to minimize this by maintaining a lower butter temperature. However, some reaction does take place at lower temperatures. This can be observed by comparing the flavor of butter with dextrose added before grinding with one prepared by adding powdered dextrose after milling and cooling. The use of sucrose, which is a nonreducing sugar, will not cause this particular flavor to develop even when the sugar is added before milling.

Dextrose monohydrate is stable below 180°F but dehydrates above that temperature. When hydrated dextrose is used in peanut butter formulations that are ground through a single stage mill, the heat evolved is sufficient to dehydrate the dextrose. The moisture transfers to the peanut protein and causes the butter to thicken considerably. Reducing the temperature of the butter will reduce its viscosity as the dextrose then rehydrates. Dextrose can form larger crystals during rehydration. The effect is similar to not having ground the dextrose. The solution lies in using anhydrous dextrose, installing a two-stage grinding system, or adding the normal dextrose hydrate in preground form to partially cooled butter after it is milled.

Moisture affects peanut flavor. An extreme example can be readily observed by stirring peanut butter into an equal amount of water. Small amounts of water such as would be obtained from corn syrup, honey, or other high moisture sweeteners alter the flavor of the butter. The moisture must be removed before the butter is packaged as it will also cause an undesirably high viscosity and pastiness. Passing the hot, high moisture butter through a deaerator strips off moisture as well as air. The flavor resulting from the temporary presence of high moisture is unique and not found in ordinary peanut butter.

Stabilizer and salt are usually added to the peanuts before grinding. The milling temperature must be sufficient to melt the stabilizer if it is added in solid form. Flaked or powdered stabilizer and salt, often premixed, can be proportioned into the peanut stream by use of a screw conveyor. Melted stabilizer can be added by gravity feed from a heated supply tank. Pumpable stabilizer-salt mix in oil suspension is also available.

Deaeration

Milling peanuts incorporates air into the butter. On being chilled, the butter will become streaked due to variable distribution of this air. Larger bubbles will have an undesirable appearance in the butter.

Deaerators are used to remove unwanted air from peanut butter. The hot butter is pumped into the top of a large closed tank under vacuum.

It flows down the tank walls for maximum surface exposure. The air boils out and is removed. The deaerated butter is then pumped to the chilling machine.

Chilling and Filling

The standard peanut butter jar was tall and thin. This was to obtain maximum heat transfer rate from hot peanut butter in the jar. The butter was filled by gravity into the jar which was then passed slowly through a chill tunnel. The chill tunnel had air circulated through it at 30°–40°F. The jars required about 15 min to pass through the tunnel. The butter was solid enough at this point to be capped, labeled, and packed in cases. Some peanut butter may still be chilled in this way.

Modern chilling of peanut butter is done by Votator A-unit or similar internal scraped surface heat exchanger. Filling equipment is usually a pressure type which can be connected directly to the Votator outlet.

Some fillers are gravity fed. In such case, the butter is not chilled to as low a temperature so that it will flow readily into the jar.

The amount and type of stabilizer is adjusted to the filling conditions required. Glycerol monostearate solidifies at a higher temperature than triglyceride stabilizers.

Peanut butter formulated with 1.8–2.0% monoglycerides can usually be filled at a higher temperature, e.g., 120°–130°F, than when other stabilizers are used. The butter is more fluid at the higher temperature and can be filled by gravity. It will set up without further chilling.

Peanut butter stabilized with cottonseed oil hardfat at 1.8–2.0% can be filled by gravity at 110°–120°F but requires a chill tunnel for final cooling. Chilling this butter to 95°–100°F requires pressure filling as it will normally be too stiff to flow at this temperature.

Rapeseed oil hardfat has a very fine grained crystal structure. It can be used at 1.6–1.8% levels. Peanut butter so stabilized can be gravity filled at 95°–105°F without further chilling (Sanders 1964).

Packaging

Most peanut butter for retail distribution is packed in glass jars. Standard sizes are 6, 12, 18, 28, and 48 oz. Some jars still have the traditional tall, narrow shape. The modern jar is squat shaped for easier removal of the butter.

Opaque polyethylene and transparent XT polymer jars and polystyrene cups with polyethylene lids have been used for packaging peanut butter (Anon. 1950; Anon. 1965; Russo 1968). Peanut butter in a glass jar has a normal shelf-life of 2 yr. Its shelf-life in plastic seems to be 9 months to

1 yr.   At the end of that time, the butter begins to develop off-flavors due to oxygen transfer through the plastic jar wall.

Bulk peanut butter is packaged in 30-lb cans, 65-lb shortening cubes, and 500-lb steel drums.

Evaluation of Quality

A number of factors are involved in the quality of peanut butter.   Certain goals must be defined.   The product is then manufactured to meet the specifications within reasonable limits.

Firmness or consistency is an important attribute.   Too stiff or dry a butter will not spread easily.   Too soft a product, especially if improperly chilled or stored, may eventually separate and form a layer of oil on the surface.   An ASTM penetrometer or Bloom consistometer as described in Chap. 1 is usually used to evaluate peanut butter firmness.

Peanut butter is prone to develop a defect known as "pull-away."   This is a condition where the butter contracts on being cooled and shrinks from the wall of the jar.   The resulting air pocket has a disagreeable appearance.   There are a number of causes for pull-away.

Peanut butter can be forced to pull-away by holding it at a sufficiently low temperature for a long enough time.   Stiff butters show this phenomenon more readily than do soft products.   In addition, pull-away is not reversible with butters which are too firm.   Soft butters will often flow back against the jar on being rewarmed.   Shipment and storage of soft peanut butter at low ambient temperatures of northern winters may result in pull-away which is never observed.   The product may recover in the warmer retail market storage room before being removed from the case.

Pull-away results when the butter loses its ability to adhere to the wall of the jar.   Jars stored in a cold warehouse and brought into the warm process area will become fogged with moisture if the relative humidity is high enough.   This moisture will accentuate pull-away.   The same lack of adhesion would occur where coatings are sprayed on the outside of jars to reduce surface marring and are inadvertently permitted to reach the inside of the jar.

Peanut butter jars are purposely designed to have no shoulder.   The break in a straight vertical wall inside the jar caused by the curvature of a shoulder would act as a starting point for pull-away.   Sloped walls are permissible.   Any bead at the rim of the jar must be above the peanut butter surface.

Most butters are formulated to be softer and more spreadable than was previously the case.   This minimizes the threat of pull-away but increases the potential for oil separation.

Peanut protein seems to have antioxidant properties. Sulfhydryls are formed from cystine, a sulfur-bearing amino acid, during roasting. These and similar compounds are the probable antioxidants which prevent peanut oil in the butter from becoming rancid. Separation of the butter to form a surface layer of free oil removes the protective protein. Such butter usually becomes rancid in a few days after separation. This accounts for the poor shelf-life of old-fashioned peanut butter. Oil separation in stabilized peanut butter is a definite defect even if the oil were not to become rancid.

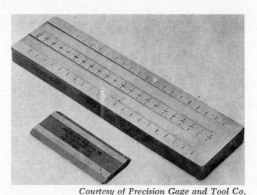

Courtesy of Precision Gage and Tool Co.

FIG. 39. FINENESS OF GRIND GAGE FOR PEANUT BUTTER

It is impractical to attempt to make a speck-free peanut butter. The sorting out of defective nuts which cause the specks cannot and need not be complete. A reasonable limit must be set on speck count. This is usually stated as a given maximum number of specks larger than a specific permitted size per square inch of peanut butter. A speck count exceeding the maximum limit usually indicates blocking of the sorter which is supposed to remove shriveled nuts.

Finely ground peanut butter should have particles with a maximum size of about 3 mils. A more coarsely ground regular butter may have 7–11 mils maximum particle size. Peanut butter fineness of grind gages are available. One such gage is shown in Figure 39. It consists of a metal bar with a sloped, flat bottom trough ground lengthwise into it. The trough is graduated in depth from 0 to 20 mils. A mass of peanut butter to be evaluated is placed at the deep end of the trough. A scraper blade is used to sweep the butter the length of the trough. Streaks begin to appear in the butter at the depth corresponding to the maximum particle size of the sample. Isolated single streaks corresponding to random, large particles are usually ignored.

Color of the butter is important. It indicates the extent of roast of the peanuts. The amount and character of butter flavor is related to darkness of roast. Too light a roast will result in a butter low in flavor content and even too high in moisture to have a soft consistency. Too dark a roast will give a burnt flavor which may also be too bitter. Color standards are available for U.S. Dept. of Agr. peanut butter. They consist of four colored plastic sticks graded from light to dark (Magnuson Engineers, San Jose, Calif.). They may be used as guides for any peanut butter. It is also possible and more practical to make up a set of house standards by setting aside jars of peanut butter which illustrate the extreme limits of acceptable color for the particular product involved.

BIBLIOGRAPHY

ANON. 1950. Peanut butter in plastic. Food Ind. 22, 2122.

ANON. 1965. Blow-molded polyethylene jar. Food Eng. 37, No. 12, 30.

ANON. 1968. Whipped peanut butter. Food Prod. Develop. 2, No. 2, 14.

AVERA, F. L. 1966. Process for making a chunk style peanut butter. U.S. Pat. 3,246,991. Apr. 19.

BAHR, G. R., and KRUMREI, W. C. 1964. Packaged peanut butter product and method of making same. U.S. Pat. 3,117,871. Jan. 14.

FREEMAN, A. F., NORRIS, N. J., and WILLICH, R. K. 1954. Peanut Butter. U.S. Dept. Agr. AIC-370, Washington, D.C.

HLAVACEK, R. G., and FREEDMAN, W. G. 1968. Low heat, 200 mesh milling provides smooth uniform product. Food Process.-Marketing 29, No. 4, 44–45.

HOLMAN, G. W., and QUIMBY, O. T. 1950. Process for preparing suspensions of solid triglyceride and liquid oil. U.S. Pat. 2,521,219. Sept. 5.

JAPISKE, C. H. 1966. Flavor improved peanut butter. U.S. Pat. 3,265,507. Aug. 9.

JENKINS, G. E. 1968. Watch guidelines to obtain peanut value, quality. Candy Ind. 131, No. 7, 7–8.

MITCHELL, P. J. 1950. Pumpable shortenings. U.S. Pat. 2,521,242. Sept. 5.

PICHEL, M. J., and WEISS, T. J. 1967. Process for preparing nut butter from soybeans. U.S. Pat. 3,346,390. Oct. 10.

RUSSO, J. 1968. Switch to plastic jars increases peanut butter sales 80%. Food Process.-Marketing 29, No. 3, 22.

SANDERS, J. H. 1964. Stabilized peanut butter. U.S. Pat. 3,129,102. Apr. 14.

TIEMSTRA, P. J. 1969. Personal communication. Chicago, Ill.

WOODROOF, J. G. 1966. Peanuts: Production, Processing, Products. Avi Publishing Co., Westport, Conn.

CHAPTER 12

Confectionery Coatings

Confectionery coatings are blends of sugar and other solids suspended in fats which are hard at room temperature but are melted or at least softened at body temperature. The best known of such coatings is chocolate. The fat in chocolate is cocoa butter; the solids, other than sugar, are cocoa and frequently milk.

Similar coatings are used for enrobing some candy bars and various cookies. The fat in that case is one of several hard butters made especially for the purpose. The product cannot legally be called chocolate. It is called a compound coating. The U.S. FDA Standards of Identity refer to it as "sweet cocoa and vegetable fat (other than cacao fat) coating."

A special group of coatings are made for covering ice cream bars and other frozen desserts. One type uses cocoa as the flavoring ingredient, the other, chocolate liquor. These are described under different standards from each other.

Pastel coatings contain no cocoa. They are artificially colored, usually light pink, green, or yellow. So-called "white chocolate" is a pastel coating formulation using cocoa butter as the suspending fat. These products are not covered by standards.

CHOCOLATE

A number of books have been written about cocoa and chocolate. Cook (1963) and Kempf (1964A) wrote treatises on chocolate alone. Chatt (1953) is on cocoa products. Other books are general texts on confectionery products (Jordan and Langwill 1946; Williams 1950; Leighton 1952; Richmond 1954). All of these books are quite detailed in the fine points of chocolate making. For example, they discuss the selection and blending of various cocoa beans for flavor. Details are given on all aspects of each processing step as well as for many formulations.

It is the purpose of this chapter to present only the basic aspects of chocolate and related products.

Standards of Identity

There are several sections of the U.S. FDA Standards of Identity devoted to chocolate. They are titled "Sweet Chocolate," "Milk Chocolate," "Skim Milk Chocolate," "Buttermilk Chocolate," "Mixed Dairy Product Chocolates," "Sweet Chocolate and Vegetable Fat (other than cacao fat),"

"Milk Chocolate and Vegetable Fat (other than cacao fat)," and "Sweet Cocoa and Vegetable Fat (other than cacao fat)." As the titles suggest, most product types differ in the milk constituent. Even the sweet chocolate Standard permits milk solids but at a level of less than 12% by weight.

Sweet chocolate must contain a minimum of 15% chocolate liquor by weight. Bittersweet chocolate contains not less than 35% chocolate liquor, milk chocolate not less than 10%.

When vegetable fats other than cocoa butter are added to chocolate liquor, they must have a melting point below that of cocoa butter in order to use the label "Chocolate." Coatings of this type can not be used as ordinary chocolate. They have too low a melting point. These coatings are for enrobing frozen dessert products. Higher melting fats may be added to chocolate liquor or to cocoa powder. In either case, the resulting product is a "cocoa" coating. The nonfat cocoa solids must not be less than 6.8% by weight of the coating.

Chocolate contains one of several combinations of sweetener. In each case the sweetener is sugar alone or blended with dextrose or corn syrup solids (40 DE minimum).

Various flavorings are permitted except for those described as imitating the flavor of chocolate, milk, or butter. The most commonly used flavors are vanillin, ethyl vanillin, heliotropin, and salt.

Specific emulsifiers are permitted in chocolate and at specified maximum limits. Lecithin, mono-, and diglycerides are allowed at 0.5% by weight each or a total of 0.5% in combination. Sorbitan monostearate may be used at 1% maximum. Polysorbate 60 may not exceed 0.5%. A combination of these, exclusive of lecithin, must not exceed 1.0% by weight of the total chocolate composition. Polysorbate 60 and sorbitan monostearate cannot be used in coatings designed for frozen desserts as they serve no useful purpose not already accomplished by lecithin or monoglycerides.

Ingredients

Cocoa Products.—Chocolate liquor is merely roasted and ground cocoa beans. The beans grow in pods on the cacao tree. The beans are picked when ripe and fermented in order to develop compounds which result in a desirable flavor on being roasted. Some low cost beans are not fermented.

Cacao beans from various parts of the world differ in ultimate flavor character, partly due to soil and climate conditions, partly to locally traditional fermenting methods. The beans are roasted either batch-wise or on a continuous basis. They are then cracked and shelled. The roasted, shelled beans are known as "nibs."

The nibs are ground to a paste or liquor. Liquor mills are made of a series of revolving stones, steel plates, or steel rolls. The liquor contains 50 to 55% fat which makes it quite fluid when melted. Liquors from different beans are usually blended for special flavor effects. Some have good aroma with mild flavor. Others lack aroma but have a heavy flavor.

Chocolate liquor is made into finished chocolate by adding sugar and milk solids. There is not a sufficient amount of fat in the liquor to prepare a satisfactory coating. Additional cocoa butter must be made available. This is done by pressing liquor in a hydraulic press. The resulting products are cocoa butter and cocoa powder.

Cocoa butter is a fragrant but almost flavorless fat which melts at 89°– 95°F. It has many polymorphic forms: alpha, beta, beta-prime, beta-double prime, and gamma (DuRoss and Knightly 1965). It is stable in the beta phase. Cocoa butter will separate into large solid crystals and liquid oil on slow cooling. After several days, the liquid portion will also solidify. The entire mass can be solidified into a hard, brittle fat on faster cooling, especially if it is agitated or worked during the chilling operation.

Cocoa powder is prepared by grinding the compressed liquor solids removed from the hydraulic press. The amount of fat remaining in the cocoa depends on the pressure and pressing time involved. Breakfast cocoa contains at least 22% fat. Medium fat cocoas range from 10 to 22% fat. Low fat cocoa has less than 10% residual fat. Cocoa coatings are usually prepared from cocoas containing 7–12% fat to minimize the potential of bloom development.

Lower fat cocoas could be prepared by solvent extraction. Cocoa powder contains very active natural antioxidants. Solvent extraction removes these compounds. The resulting cocoa is not desirable for coating manufacture due to this lack of antioxidant property (Minifie 1968A).

Cocoa is a source of lipolytic enzyme. The lipase is inactivated by heating the cocoa to over 240°F (Minifie 1968A). This is essential for cocoa powders to be used in compound coatings made with lauric acid fats.

Dutch process cocoa is prepared by treating cocoa nibs with alkali. Potassium carbonate is the most commonly used alkali. One process calls for mixing 2.5 lb potassium carbonate per 100 lb nibs. The carbonate is added as a 5–12.5% aqueous solution and the entire mass is heated to 175°–185°F for 1 hr before the nibs are ground into liquor (Minifie 1968B). A number of theories have been advanced to explain what takes place during alkalization. From a practical standpoint, the chocolate develops a characteristic mellow flavor and a dark color.

Cocoa is practically odorless. Any noticeable odor usually comes from the residual cocoa butter. The odor level is proportional to the fat content of the cocoa. The flavor of chocolate comes entirely from the cocoa solids. It is primarily bitter and astringent with a characteristic roasted flavor. A similarity exists between the flavors of cocoa and various other roasted or heated food materials, coffee, peanuts, cereal grains, and malted milk.

Milk Products.—Although a number of milk products are permitted in chocolate manufacture, those most frequently used are milk solids nonfat and whole dried milk powder. Fluid and condensed milk contain 50–90% water which must be removed during some phase of processing. While it is possible to achieve special flavor effects by drying milk products in combination with chocolate liquor, it is much less costly to remove water from milk in the dairy plant.

Milk is dried by one of two basic methods, by drum or by spray. Drum drying, especially on atmospheric rolls, is a high heat process. Milk is allowed to flow into a trough formed by two steam heated rolls revolving in opposite directions. The rolls pick up a thin film of milk which is dried by the steam. Doctor blades scrape the dried milk from the rolls before that portion of the roll reaches the milk supply trough. The milk is in the form of fine flakes. Atmospheric drying yields a dark, slightly caramelized product with a strong scorched milk flavor. Vacuum drying yields a light colored, almost bland powder.

Roller or drum-dried whole milk is a relatively oily powder. The butterfat in this type of product is smeared through it in large masses. This is especially important for wet conched or unconched chocolate where the butterfat needs to be readily available.

Spray-dried milk is more commonly produced. It is prepared by pre-condensing milk and spraying it into a large chamber along with a forced hot air draft. The fine spray dries quickly and is removed from the spray chamber as powdered milk solids. Spray drying is a low heat process. If high heat powder is required, the heat must be applied when the milk is in fluid form.

Whole milk is homogenized before or during spraying. Butterfat in the whole milk powder particle is still in a finely dispersed state. It is not entirely available unless the chocolate mix containing such powder is dry conched or worked in equivalent fashion. This is important as the flow characteristics of finished chocolate are dependent on the amount of freely available fat and not on total fat content.

Milk contains active lipase. Low temperature pasteurization, e.g., at 155°F, will inactivate lipase only temporarily. Use of such milk in chocolate products will result in slow formation of free fatty acids from any fats present due to lipase activity. If whole milk, cream, or butter is incorporated in chocolate containing low heat spray powder, the butterfat will hydrolyze to release butyric and other short chained fatty acids. These acids blend well with the flavor of chocolate. This is the basis of the unique flavor of American milk chocolate.

Pasteurization temperature in excess of 170°F results in permanent inactivation of milk lipase. Fats in chocolate made from milk products in which the lipase has been inactivated will normally not hydrolyze. Chocolate is more delicately flavored in the absence of butyric acid. Swiss style milk chocolate is this type of product.

Miscellaneous Ingredients

The other vegetable fats referred to in the Standards for chocolate for enrobing frozen dessert are usually unhydrogenated coconut oil, blends of hydrogenated and unhydrogenated coconut oil and occasionally palm kernel oil. Coconut and palm kernel oils contain lauric acid as flavorless glycerol esters. Hydrolysis of these oils to yield free lauric acid causes formation of a strongly disagreeable soapy flavor. Active lipase must be excluded from all products containing lauric acid fats. Chocolate, cocoa, nut meats, and milk products used in conjunction with coconut or palm kernel oil must be processed at a sufficiently high temperature to permanently inactivate their lipases.

Salt and sugar are crystalline compounds. These materials can cause excessive wear on refining rolls. Salt is especially hard. Sugar and salt should be obtained in as small a crystal as is economically feasible.

Formulation

Typical formulas for various types of coatings are given in Table 26. The amounts of chocolate liquor, milk powder, and sugar are subject to considerable variation depending on the flavors desired. The level of cocoa butter to be added is usually adjusted to yield a total fat content of 31–33% for enrobing chocolate. This is the amount required to give a flowable mass.

Lecithin reduces coating viscosity. It is added at a level corresponding to 1% of the fat content. It seems to reach a maximum effect at about this concentration. Additional lecithin may reduce viscosity further but not sufficiently so to warrant using more.

Sorbitan monostearate and polysorbate 60 functionality was studied by DuRoss and Knightly (1965) among others. They arrived at the conclusion that the optimum levels of usage for these compounds in chocolate was 1% of a blend of 60% sorbitan ester and 40% polysorbate. The combination resulted in chocolate with improved gloss and bloom resistance.

TABLE 26

TYPICAL CONFECTIONERS' COATINGS

| | Coating Type, Weight % | | | |
Ingredient	Milk Chocolate	Sweet Chocolate	Compounds	Pastel
Sugar	47.8	48.3	48.0	9.8
Chocolate liquor	16.0	31.0	...	...
Cocoa powder (10% fat)	...	...	8.0	...
Cocoa butter	18.2	19.2	...	...
Hard butter	...	...	30.7	31.7
Whole milk powder (26% fat)	16.5	...	...	...
Msnf	...	...	11.8	18.0
Lecithin	0.3	0.3	0.3	0.3
Sorbitan monostearate	0.6	0.6	0.6	0.6
Polysorbate 60	0.4	0.4	0.4	0.4
Salt	0.2	0.2	0.2	0.2

Flavor: vanillin, ethyl vanillin, heliotropin
Food colors are added to pastel coatings.

Vanillin and related flavors are added in amounts subject to personal taste. A range of 0.03–0.06% is common. The salt level is also based on personal judgment. It is not necessary to add salt, but a small amount seems to enhance overall flavor of the chocolate.

Coating Manufacture

Equipment.—Various special types of equipment have been developed for chocolate processing. Many of these pieces of apparatus have French names since much of the original development was done by French-speaking people. The term couverture is still used for chocolate coating.

The chocolate manufacturing process begins in a melangeur or mixer. Any medium duty mixer with intensive mixing action can be used. A muller most closely resembles the original melangeur but planetary mixers are also satisfactory.

Refining rolls are used to reduce the particle size of the solids suspended in the cocoa butter. Figure 40 shows a typical 5 roll chocolate refiner. The particle size desired for a good grade of chocolate is approximately one mil. The action of the roll mill is to rub the particles between the heavy rolls. This action is accomplished by having each roll turn in the opposite direction to the preceding roll and at twice its rpm. The end roll in a 5 roll mill turns at 16 times the rate of the first roll.

The conche is a mixer of special design (Minifie 1969). The term means shell and may refer to the shape of the original machine. Kempf (1964B) prefers the word "conge" which is the phonetic spelling for the way "conche" is pronounced. The early conche was a stone pot with a stone roll which plodded laterally through the chocolate mass. The action of the roll was to break up the sugar and cocoa solids, spreading the fat over the surface. The mixture contained lecithin and was fluid at the start. As conching progressed, the chocolate particles became finer and the fat layer became thinner. The mass became more viscous. Cocoa butter was added periodically to thin the mass.

Courtesy of The Buhler Corp.

FIG. 40. FIVE ROLL CHOCOLATE REFINER

Roll mill has hydraulic control of roll pressure. Water temperature of rolls is controlled automatically.

The rotary dry conche (Fig. 41) is a recent development. The working blades revolve about a vertical shaft. The conche is jacketed to heat its contents although working action of the blades can eventually cause the temperature of the mix to be self-sustaining. If necessary, the jacket can be filled with cooling water. The conche bowl is tilted to increase surface exposure of the contents. Another conche design provides a hot air blower to aid in removal of moisture and various volatile compounds from the chocolate mass.

The rotary conche made dry conching feasible. Here the chocolate mix is started dry and conched to a fluid paste. Dry-conched chocolate is formulated with less cocoa butter than is feasible with the wet-conched product. This results in a lower ingredient cost.

Process.—Chocolate coating is prepared by mixing the dry ingredients, chocolate liquor, and part of the cocoa butter in a melangeur. The amount of cocoa butter is just sufficient to make the mix greasy. No lecithin or other emulsifier is incorporated at this time. The thoroughly mixed ingredients are passed through the refiner to reduce the particles to the desired size. The chocolate is fluid at this stage but soon hardens to a powdery mass. This material is then put into a rotary conche.

*Courtesy of Hermann Bauermeister
Maschinenfabrik, G.m.b.H.*

Fig. 41. Rotary Dry Conche for Chocolate

As conching progresses, the powder becomes progressively more oily. It forms small beads which form into large balls in time. The balls coalesce into a heavy paste and later a thick fluid. Hot air is blown over the mass. The chocolate temperature is adjusted to a desired level. Sweet chocolate is usually conched at 130°–200°F. Milk chocolate may caramelize excessively and develop off-flavors at that high a temperature range. Milk chocolate is conched at 110°–135°F.

The balance of the cocoa butter is worked into the chocolate after it becomes fluid. This is followed by the lecithin and other emulsifiers. The paste becomes quite fluid after addition of the lecithin.

The conche is used to break up the cocoa particles which were flattened by the refiner. Conching also works the finely distributed butterfat globules out of the entrapping spray-dried whole milk solids. If too much cocoa butter is added before conching and especially if lecithin is added prematurely, the chocolate will immediately become too low in viscosity for the conche to break up the particles by physical action. The chocolate will retain this insufficiently low viscosity even after long additional working by the conche.

Removal of moisture, volatile acids, ketones, aldehydes, and other compounds during conching smoothens and mellows the natural chocolate flavor. Vanilla and other artificial flavors are usually added at the end of conching so as not to lose them through volatilization.

The rotary conche requires 6–16 hr to reach the point of fluidity of the original mix. Additional conching of 8–12 hr is needed after the remaining cocoa butter and lecithin are added. The old style conche required several days to produce chocolate of a quality which would satisfy the old style chocolate maker. In contrast, a continuous conche requiring 1 hr for complete conching has been developed and installed in an English chocolate factory (Minifie 1969).

After conching, the coating is usually poured into pans to harden. The solidified slabs are depanned, packaged, and shipped to candy manufacturers for their use.

Enrobing Equipment and Process

Candy Bars and Pieces.—An enrober is a device which deposits fluid coating on candy centers to be covered. The centers, nougats, nut clusters, caramels, etc., are first deposited on a continuous belt which passes through the enrober. The pieces should be between 75° and 78°F for proper enrobing.

The chocolate coating is melted in a holding tank which is the main supply for the enrobing operation. The supply tank feeds the melted coating to a small seeding pot which mixes seed or precooled chocolate with the freshly melted stock. The coating flow rate and viscosity are adjusted to correspond to the throughput rate of the centers. This applies a definite thickness of coating to the enrobed piece.

Coating is melted at less than 120°F for sweet and less than 110°F for milk chocolate in order not to cause flavor problems from overheating. The enrober supply tank is held at 92°–95°F without seeding.

Seed is prepared by cooling chocolate to a "mush" stage at which point it thickens to a heavy paste. This indicates the presence of a large quantity of beta crystals in the cocoa butter. The mush is heated to just barely melt the crystalline mass if the coating is to be used directly for dipping or hand enrobing. Otherwise, unseeded molten coating stock is mixed with the mush. Old chocolate or cold stock left overnight in the feed tank-seeding pot is often carefully remelted for use as seed.

The flow rate of unseeded stock to the enrober feed tank is adjusted so that 2–4% seed is mixed with fresh coating. This should result in 0.5–1% beta crystals in the mix. DuRoss and Knightly (1965) recommend the higher level of beta crystals for best results when sorbitan-polysorbate esters are used. The recommended temperatures for seeded mix in the enrober supply tank are 88°–89°F for sweet and 86°–87°F for milk chocolate. The supply tank has a water jacket to control this temperature.

Soft creams are often covered by casting the coating into molds. The hollow cast shells are hardened, filled with the cream, and covered with a thin layer of melted, seeded coating. The final covering becomes the bottom of the piece. Chocolate bars are also cast in molds.

If the chocolate is not properly tempered, it will not set up properly. Cast pieces must release easily from the mold. They should shrink slightly in the cooling tunnel but will not do so with poor temper. Improperly tempered chocolate, whether enrobed or cast, will also have poor gloss and resistance to bloom.

Various devices have been developed to obtain uniform and proper temper. Many of these are heat exchanger systems (Bolanowski 1967; Schuemann 1968). One type consists of a heating section followed by a cooling section and finally by another heating unit. Mitchell (1968) suggests cooling to 3°–5°F below output temperature. A variation of the continuous system calls for returning part of the chilled coating to the stream of chocolate entering the first heating stage. This recycling of a portion of the output material is common practice in the crystallization of other products. Mitchell (1968) does not recommend recycling as it may cause overtempering and consequently too viscous a coating.

Other tempering devices are based on the mechanical working principles outlined by Feuge et al. (1962). Intense physical working of solid or supercooled fluid but untempered chocolate was found to transform unstable fat crystals to the stable beta form in a relatively short time (Feuge and Landmann 1965). Remelting the worked solid chocolate just sufficiently to cast it into a mold and resolidifying it in a cooling tunnel resulted in a properly tempered bar. Unworked chocolate did not release from the mold. Bolanowski (1967) refers to shock chilling and working of the coating in a Votator system.

The enrobed or cast chocolate pieces are finished by passing them through a cooling tunnel on a continuous belt. The cooling tunnel is about 60 ft in length. Cold air is forced through the tunnel in one or more stages. Air temperature may be 60°–68°F at the tunnel entrance and 50°–60°F at the exit. Some tunnels have a bank of infrared heaters or warm air flow in a center section to partially remelt the chocolate. This is supposed to result in an improved gloss (Cook 1963).

After exiting from the cooling tunnel, the enrobed pieces or depanned bars are wrapped, boxed, and placed in storage at 65°–68°F with a 55% maximum relative humidity.

Frozen Dessert Coating.—Ice cream, ice milk, and mellorine bars which have been frozen on a stick are coated by dipping. The frozen but still fluid dessert mix is pumped into a mold and covered with a slotted plate holding the sticks in an upright position. The mold is passed through a cold chamber or bath to solidify the bars. The cover of the mold is removed. A rack of clamps grasps the rows of frozen bars by the protruding sticks. The mold is then pulled away.

The uncoated bars are suspended from the rack and moved over a bath of melted coating. The coating temperature is usually no more than 95°F. It has a low melting point, often 80°–85°F. The bars are immersed for 1 or 2 sec, and drained briefly. The bars are passed through a cold tunnel and wrapped.

The coating should set up firmly in 7–11 sec. This is the length of time from dipping to bagging in high speed coating machines. The coating viscosity should be low to prevent excessive thickness of coating on the bar. No tempering is required for this type of product.

Coatings for frozen dessert are soft at room temperature. They are packed in 60-lb pails or 500-lb drums and are referred to as "pail" coatings.

Evaluation of Quality

Temper.—Much effort is expended in obtaining the proper temper for chocolate. It is, therefore, essential to be able to measure the degree of temper in the product before it is used for casting or enrobing. DuRoss and Knightly (1965) have devised a viscosimeter for this purpose. They observed that the viscosity of tempered chocolate was related to the proportion of crystalline fats therein. A modified Brabender Visco-corder is available for making such measurements (Anon. 1968).

Most observers have agreed that chocolate must exist in the beta crystalline form to have a good gloss and proper temper. Feuge *et al.* (1962) have obtained X-ray diffraction patterns to prove this point. Wonsiewicz and Paulicka (1967) have disputed this. Their X-ray diffraction patterns show tempered chocolate to exist in the beta-prime form. They were unable to obtain beta crystals. Their theory is that tempering does not change polymorphic form (beta-prime to beta) but causes fractional crystallization of the chocolate fat. Gloss is a result of supercooling fats. On that basis, fat bloom may result from the formation of beta crystals on the chocolate surface during heating and subsequent cooling of the chocolate.

Fat Bloom.—Bloom is a surface defect whereby initial gloss of the chocolate is first lost and is then replaced by a white haze. In contrast to the theories involving fat crystal transformation as a cause of bloom formation, DuRoss and Knightly (1965) suspect that bloom results from the migration of melted chocolate fat to the nonfat solids on the chocolate surface during heating. Recrystallization of free fats on the surface on recooling forms the bloom. No polymorphic crystal changes occur here since beta crystals merely melt and resolidify as such. Sorbitan monostearate supposedly inhibits bloom formation by retarding migration of the melted fats to the chocolate surface.

Resistance to bloom is usually measured by cycling the test chocolate between high and low temperatures. High temperatures may be 80°–85°F and low temperatures 60°–65°F. Time at each temperature may be 12–24 hr. The relative humidity should be low, e.g., 25%, to prevent sweating of the chocolate on being moved from low to high temperature. The number of cycles required to obtain observable bloom is related to bloom resistance. DuRoss and Knightly (1965) report that a chocolate containing 1% of a blend of 60% sorbitan monostearate and 40% polysorbate 60 requires 15 to 24 cycles to reach first bloom. A control without these surfactants bloomed after only five cycles.

Poor temper can cause chocolate bloom even before the product leaves the cooling tunnel. If centers to be enrobed are too warm, they will melt seed crystals and thereby destroy the temper. Too cold a center will cause the coating to solidify too quickly. This will prevent the necessary transformation of unstable to stable crystals and also result in lack of temper. Centers at 75°–78°F promote proper temper by slight cooling of the newly deposited coating.

Sugar Bloom.—Excessive moisture can cause sugar in the chocolate to partially dissolve and recrystallize. This formation of visible sugar crystals is referred to as "sugar bloom." There are several potential sources of excessive moisture. The most common cause of sugar bloom is in moving cold chocolate into a warm, humid area. Moisture will condense on the chocolate.

Occasionally, cold storage under high relative humidity (over 55%) will cause sugar bloom. High residual moisture in the chocolate or enrobing a high moisture center with chocolate containing too little fat will also result in this defect.

Cooling and Heating Curves.—Cooling curves have been used to determine the comparative merits of various hard butters in relation to cocoa butter (Fig. 42). These curves were obtained by observing the temperature of the sample at intervals during relatively rapid cooling. Cocoa butter sets up more slowly and at a lower temperature than do the various hard butters. The results cannot show the potential of a fat to be tempered or to require temper, however.

Johnston and Price (1968) have used a differential scanning calorimeter to measure the melting curves of cocoa butter. These curves have been interpreted to show percent solid fats at any given temperature. The advantage in melting curves is that the butter, chocolate liquor, or finished coating can be evaluated from an untempered, partially tempered, or tem-

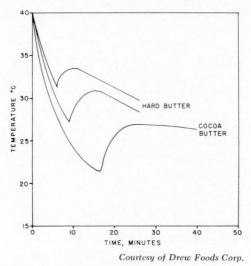

FIG. 42.   COOLING CURVES OF COCOA BUTTER
AND TYPICAL HARD BUTTERS

pered state. Figure 43 shows curves obtained for cocoa butter pretreated by 3 different methods. Only fully tempered cocoa butter was found to contain 100% solid fat at room temperature.

Fineness of Grind.—The Chocolate Manufacturers Association has adopted a Precision gage for measuring particle size in chocolate. The gage is similar to the one described for evaluation of peanut butter. The chocolate gage has 2 channels, one 0–105 μ, the other 80–185 μ in depth. The sample is mixed with an equal volume of warm mineral or coconut oil so that it will remain fluid for ease in making the measurement (Kribben 1969).

Scraping of the sample along the channels reveals the presence of an occasional outsized particle among the mass of smaller particles. One popular method for determining particle size was to place a drop of oil-diluted chocolate in a micrometer caliper to measure the size of the largest particle. If an outsized particle chanced to fall between the caliper jaws, a false reading would be obtained. The Precision gage would obviously offer a more reproducible measurement.

COMPOUND COATINGS

Compound coatings are covered by Standards of Identity under the heading "Sweet cocoa and vegetable fat (other than cacao fat) coating." Two basic types of compound coatings are possible. One is similar to chocolate in that chocolate liquor is used as the source of cocoa. Hard

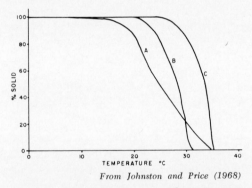

From Johnston and Price (1968)

Fig. 43. Melting Curves of Cocoa Butter

Sample was tempered in ways to give (A) mixed crystal forms, (B) metastable beta-prime crystals, and (C) stable beta crystals.

Reprinted from *The Manufacturing Confectioner*

butter, as the vegetable fat is called, is used in place of added cocoa butter in the coating formula. The proportion of hard butter to cocoa butter is 2 to 3:1. The hard butter must be compatible with cocoa butter within this ratio range. The cocoa butter portion of the fat blend contributes to the overall flavor of the coating.

Cocoa coatings are made with hard butter as the sole source of fat. This hard butter need not be compatible with cocoa butter. A formula such as is given in Table 26 contains only 2.5% cocoa butter on a total fat basis if the cocoa has 10% fat content. This amount has no significant effect on melting characteristics of the hard butter. Cocoa coatings are flavored with vanilla type compounds and the cocoa itself. They lack the fragrance of chocolate since hard butters are completely bland in odor and flavor.

Hard butters are divided into two major classes, the lauric acid fats and the domestic or nonlauric fats.

Lauric Acid Hard Butters

Compatibility with cocoa butter requires that the hard butter be similar to cocoa butter in composition (Feuge 1964). Lauric acid fats are not similar to cocoa butter and are incompatible with it.

Lauric hard butters are prepared by several methods. One of the earliest was to chill palm kernel oil and press it. The stearine portion has melting characteristics required of a hard butter. It melts at 90°–93°F, about the same melting point as cocoa butter. A more recent variation of this is to fractionate palm kernel oil from solvent. Roylance (1960) prepared a hard butter by hydrogenating palm kernel oil and interesterifying it before fractionating it from solvent. The product had a melting point around 100°F.

Wonsiewicz and Paulicka (1967) observed that fractionated lauric hard butters became more beta-prime in crystal structure on being tempered. This tended to make them more stable and bloom resistant.

Teasdale and Helmel (1965) describe the production of a series of hard butters ranging from 95° to 115°F in Wiley melting point. They fully hydrogenated palm kernel oil. A portion of this material (15–85%) was interesterified and blended back with the balance of the uninteresterified, hydrogenated palm kernal oil. The melting point of the blend increased with increasing proportion of the uninteresterified component.

Cochran and McGee (1955) and Cochran and Ott (1957) prepared a series of hard butters by interesterifying fully hydrogenated coconut or palm kernel oil with fully hydrogenated cottonseed or other nonlauric oil. Melting points of 102°–118°F are feasible with this method. The melting point obtained is proportional to the amount of hydrogenated cottonseed oil in the starting material. Kidger (1968) reports that fully hydrogenated rapeseed or marine oil in place of the cottonseed oil component gives an unusual gloss and gloss retention to finished coatings prepared from such hard butters. The high level of 20 and 22 carbon fatty acids from rapeseed and marine oils seems to contribute a particularly fine-grained crystal structure with good stability.

The higher melting points obtained from some of these hard butters is desirable for candy bar and cookie enrobing. Chocolate has too low a melting point to be generally useful, especially in the summer. It is not necessary for coatings to melt completely in the mouth. They are masticated with other components of the confection which do not melt or dissolve. Lack of melting often goes unnoticed by the consumer. Hard butters for preparation of candy bar and cookie coatings are available in stepwise Wiley melting point ranges, the most common being 94°–96°, 101°–103°, 107°–109°, 112°–114°, and 117°–119°F.

Coatings prepared from these hard butters are tempered and enrobed in the same manner as chocolate except for the temperatures used. Seeding should be carried out 3° to 5° below the melting point of the coating.

Biscuit and cracker coatings are lower in viscosity than candy coatings. The enrober is equipped with a blower to literally air-blow excess coating from the enrobed piece. The resulting layer of coating is as thin as is practical for consumer acceptance.

Coconut oil with or without added hydrogenated coconut oil is used for making cocoa coatings for frozen dessert use. These are more commonly used than the chocolate with vegetable fat coatings previously described.

Any material with lipolytic activity must be excluded from lauric acid fat hard butters and coatings containing them to prevent development of soapy off-flavors. Lauric fat hard butters are often packaged with 0.1% lecithin added as a moisture scavenger. This reduces the potential of developing hydrolytic rancidity in the stored fat.

Domestic Fat Hard Butters

Various fats have been fractionated from solvent to yield disaturated triglycerides which have similar melting characteristics to that of cocoa butter. A number of types have been described in patent and other literature.

Feuge et al. (1958) describe fractions prepared from esterification of oleic, palmitic, and stearic acids with glycerin. He also prepared a fraction from an interesterified blend of 70% fully hydrogenated cottonseed oil and 30% liquid olive oil as a source of oleic acid. These fractions were not entirely compatible with cocoa butter. The major triglyceride in cocoa butter is 2-oleopalmitostearin. The presence of oleic acid in the 2 position of glycerin requires that a perfect triglyceride substitute for cocoa butter also have oleic acid in this position. The fractions prepared by Feuge et al. (1958) had oleic acid distributed randomly among all three glycerin carbon positions. Thus the fractions acted as impurities in cocoa butter and depressed its melting point.

Various natural fats have been fractionated. These include palm oil (Best et al. 1960), lard (Crossley 1958A), and tallow (Crossley 1958B). These fractions may be mixed in use. Borneo tallow may also be added to palm oil fraction to make it harder in texture. Crossley and Paul (1959) fractionated partially hydrogenated palm oil. Wissebach (1960) separated out the softer monosaturated triglycerides and hydrogenated them to cocoa butter hardness. These fats all have hard butter attributes. They can be used in cocoa coatings. Compatibility with cocoa butter for preparation of chocolate liquor coatings is another matter. Some of the above fractions may be compatible, others may not.

Two hard butters are claimed to be compatible with cocoa butter. One of these (Cochran *et al.* 1961) is prepared by partial hydrogenation of soybean and cottonseed oils to an iodine value of about 60. The linoleic

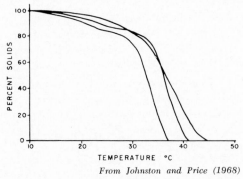

From Johnston and Price (1968)

Fig. 44. Melting Curves of Typical Hard
Butters

Reprinted from *The Manufacturing Confectioner*

acid in these oils is converted to iso-oleic acid. This *trans*-isomer behaves somewhat like palmitic acid. The fat is fractionated from solvent to yield a cocoa butter-like fat with 40–50% *trans*-acid content and approximately 100°F melting point.

The fraction obtained is harder than cocoa butter. A blend of this fat with cocoa butter, however, results in a mixture which has melting characteristics similar to those of cocoa butter by itself. A mixture of two dissimilar fats gives a softer product than would be expected. In this case, the calculated hardness of the blend would seem to be excessive. Softening by blending the fats results in an actual hardness which has been reduced to a desirable level. The amount of softening would depend on the relative proportions of cocoa butter and the substitute.

Another fat which is claimed to be compatible with cocoa butter is prepared by blending palm oil with Shea butter (Best *et al.* 1961). The blend is fractionated from solvent to yield a disaturated triglyceride. It is supposed to be able to replace 25–50% of the cocoa butter in a chocolate formula.

Many of these fractions crystallize in the beta-prime phase. It is claimed that coatings prepared from them can be enrobed or cast without being tempered. If tempering is required, the coatings are at least less sensitive to variation in tempering conditions than is chocolate.

Johnston and Price (1968) have measured melting curves for three commercially available nonlauric hard butters (Fig. 44). They all had higher solids and melting points than cocoa butter. They also showed no differences in behavior on being subjected to different tempering conditions. Cocoa butter varied widely with change in manner of preparation. This seems to bear out the claim that coatings prepared from at least these cocoa butter substitutes require no special tempering conditions.

The nonlauric hard butters can be used with low temperature milk powders and other lipase containing materials. They do not develop soapy or other off-flavors on hydrolysis. Their fatty acid composition is similar to cocoa butter although they may be dissimilar to cocoa butter in distribution of those fatty acids on the glycerin molecule.

PASTEL COATINGS

The colorful pastel coatings are used to enrobe cookies and other confections. Table 26 shows a typical formula for such coatings. If cocoa butter is used as the hard butter, the product will have the fragrance of chocolate but not the flavor. This material is sold as "white chocolate." Many persons who are allergic to chocolate and cocoa products seem to have no reaction to white chocolate.

Palm kernel oil stearine obtained by pressing or by solvent fractionation is the most commonly used hard butter for preparation of pastel coatings. This stearine has a lower melting point than most hard butters, i.e., 90°–93°F. The nature of pastel coatings makes this an acceptable property. The high level of milk solids gives pastel coatings different attributes from those of cocoa coatings where the milk powder content is reduced. Milk powder is more porous and spongy than cocoa powder and contributes a different texture and mouth-feel to the coatings. Hard butters which are satisfactory for cocoa coatings would be too hard and crumbly for pastel coating work.

BIBLIOGRAPHY

Anon. 1968. Modified recording viscometer may permit objective test for chocolate temper. Food Prod. Develop. 2, No. 4, 89, 94.
Best, R. L. *et al.* 1960. Fats resembling cocoa butter. Brit. Pat. 827,172. Feb. 3.

BEST, R. L. *et al.* 1961. Cocoa butter substitute. U.S. Pat. 3,012,891. Dec. 12.

BOLANOWSKI, J. P. 1967. Tempering the better way. Food Eng. 39. No. 6, 57–61.

CHATT, E. M. 1953. Cocoa. Interscience Publishers, New York.

COCHRAN, W. M., and McGEE, R. F. 1955. Hard butter. U.S. Pat. 2,726,158. Dec. 6.

COCHRAN, W. M., and OTT, M. L. 1957. Process for preparing hard butter. U.S. Pat. 2,783, 151. Feb. 26.

COCHRAN, W. M., OTT, M. L., WONSIEWICZ, B. R., and ZWOLANEK, T. J. 1961. Domestic oil hard butters, coatings thereof and process for preparing said butter. U.S. Pat. 2,972,541. Feb. 21.

COOK, L. R. 1963. Chocolate Production and Use. Magazines for Industry, New York.

CROSSLEY, A. 1958A. Fats resembling cocoa butter. German Pat. 1,030,159. May 14. Also 1960. Brit. Pat. 841,316. July 13.

CROSSLEY, A. 1958B. Fats resembling cocoa butter. German Pat. 1,030,160. May 14. Also 1960. Brit. Pat. 841,317. July 13.

CROSSLEY, A., and PAUL, S. 1959. Cocoa butter-like fat. German Pat. 1,070,908. Dec. 10.

DuROSS, J. W., and KNIGHTLY, W. H. 1965. Relationship of sorbitan monostearate and polysorbate 60 to bloom resistance in properly tempered chocolate. Mfg. Confectioner 45, No. 7, 50, 52, 54, 56.

FEUGE, R. O. 1964. Confectionery fats: Their current status and potential market. J. Am. Oil Chemists' Soc. 41, 4, 26, 30, 63.

FEUGE, R. O., and LANDMANN, W. 1965. Process of preparing stable triglycerides of fat forming acids. U.S. Pat. 3,170,799. Jan. 23.

FEUGE, R. O., LANDMANN, W., MITCHAM, D., and LOVEGREN, N. V. 1962. Tempering triglycerides by mechanical working. J. Am. Oil Chemists' Soc. 39, 310–313.

FEUGE, R. O., LOVEGREN, N. V., and COSLER, H. B. 1958. Cocoa butter-like fats from domestic oils. J. Am. Oil Chemists' Soc. 35, 194–199.

JOHNSTON, G. M., and PRICE, E. F. 1968. The melting properties of cocoa butter and other coating fats. Mfg. Confectioner 48, No. 7, 25–27.

JORDAN, S., LANGWILL, K. E. 1946. Confectionery Analysis and Composition. Manufacturing Confectioner Publishing Co., Oak Park, Ill.

KEMPF, N. W. 1964A. The Technology of Chocolate. Manufacturing Confectioner Publishing Co., Oak Park, Ill.

KEMPF, N. W. 1964B. What has happened to the conge? Mfg. Confectioner 44, No. 2, 63–65.

KIDGER, D. R 1968. Compound coatings. U.S. Pat. 3,361,568. Jan. 9.

KRIBBEN, B. D. 1969. Private communication. Chicago, Ill.

LEIGHTON, A. E. 1952. A Textbook of Candy Making. Manufacturing Confectioner Publishing Co., Oak Park, Ill.

MINIFIE, B. W. 1968A. Special cocoas, their manufacture and uses. Mfg. Confectioner 48, No. 6, 48–53, 62.

MINIFIE, B. W. 1968B. Review of methods of producing cocoa for industry. Candy Ind. 131, No. 8, 39, 40, 46, 48, 50, 56.

MINIFIE, B. W. 1969. Old time conching method undergoes new principle. Candy Ind. 133, No. 8, 5, 9, 58, 61, 71.

MITCHELL, D. G. 1968. Practical chocolate tempering. Mfg. Confectioner 48, No. 6, 76, 78, 80–81.

PRATT, C. D. (Editor). 1970. Proceedings: Twenty Years of Confectionery and Chocolate Progress. Avi Publishing Co., Westport, Conn.

RICHMOND, W. 1954. Choice Confections. Manufacturing Confectioner Publishing Co., Oak Park, Ill.

ROYLANCE, A. 1960. Hard butter. U.S. Pat. 2,928, 745. Mar. 15.

SCHUEMANN, H. W. 1968. Chocolate tempering and chocolate tempering devices. Mfg. Confectioner 48, No. 5, 39–42, 59.

TEASDALE, B. F., and HELMEL, G. A. 1965. Process for the production of edible fats. U.S. Pat. 3,174,868. Mar. 23.

WILLIAMS, C. T. 1950. Chocolate and Confectionery. Leonard Hill, London.

WISSEBACH, H. 1960. Edible fats. U.S. Pat. 2,942,984. June 28.

WONSIEWICZ, B. R., and PAULICKA, F. R. 1967. Crystalline behavior of fats in confectionery coatings. Mfg. Confectioner 47, No. 5, 88, 90, 92–93.

CHAPTER 13

Imitation Dairy Products

There are two types of dairy-like products containing fats other than butterfat. "Filled" products are made essentially from defatted milk, usually milk solids nonfat (msnf) or fluid skim milk, and a vegetable oil or fat. "Imitation" dairy products contain no milk as such (Hetrick 1969). In most cases, imitation dairy products are made from sodium caseinate, the major protein derived from milk sources. Sodium soy proteinate has also been used in place of isolated milk protein.

The most important quality required by the oil component used in filled or imitation dairy products is blandness. Oils which seem bland as oils may not be sufficiently so in dairy products. Emulsification of the fat system exposes it to the tongue in extremely large surface areas. Traces of flavor compounds become accentuated if present. Deodorization of fats for dairy product use must be of a high order of efficiency.

The formulation of filled milk products is relatively simple. Whole milk or cream is replaced by skim milk or buttermilk which has been homogenized with substitutes for butterfat. Monoglycerides are added to aid in emulsification of the fat.

Filled milk contains about 3.5% fat and requires about 0.25% monoglycerides. The fat component may be a liquid vegetable oil, a low melting lauric hard butter, or a partially hydrogenated domestic oil (Ryberg 1968). The melting point of any fat or oil used should be below body temperature to avoid greasy mouth-feel.

Mellorines are essentially filled dairy products resembling ice milk or ice cream. Miller and Ziemba (1966) give a series of formulations prepared with 4–12% fat. These are given in Table 27. They recommend fats with a melting point range of 88°–92°F. Melting points up to about 105°F may be used if the fat is prepared by partial hydrogenation or by blending of hard and soft margarine oil components. Higher melting points or the presence of fully hydrogenated fats at lower melting points will cause greasiness in the mouth. Miller and Ziemba (1966) also point out that a mellorine fat must have sufficient solids at 10°–20°F to give proper texture and consistency for spooning the frozen product. The stabilizer used may be one of several vegetable gums. Mono- and diglycerides with or without polysorbate 60 are used as emulsifiers.

Simulated sour cream is another filled milk product (Loter 1967). Table 28 shows a formulation given by Miller and Ziemba (1966). Hard

214

TABLE 27

MELLORINE FORMULATIONS

Ingredient	Weight %				
Fat	4.0	6.0	8.0	10.0	12 0
Msnf	13.0	12.5	12.0	12.0	10.0
Sugar	12.0	12.0	12.0	12.0	12.0
Corn syrup solids	6.0	6.0	6.0	5.0	5.0
Stabilizer-emulsifier	0.5	0.5	0.5	0.5	0.5
Water	64.5	63.0	61.5	60.5	60.5
Color, flavor					

Source: Miller and Ziemba (1966).

TABLE 28

SIMULATED SOUR CREAM CHIP DIP

Ingredient	Weight %
Msnf	10.0
Vegetable gum	1.8
Fat	14.0
Water	72.7
Citric or lactic acid	to pH 4.4–4.7
Margarine flavor	

Source: Miller and Ziemba (1966).

TABLE 29

COFFEE WHITENER FORMULATIONS

Ingredient	Weight %	
	Liquid	Powder
Sugar	1.0–3.0	
Corn syrup solids (42 DE)	1.5–3.0	55.0–60.0
Fat	3.0–18.0	35.0–40.0
Sodium caseinate	1.0–3.0	4.5–5.5
Mono-, diglycerides	0.3–0.5	0.2–0.5[1]
Carrageenan	0.1–0.2	
Dipotassium phosphate[2]	0.1–0.3	1.2–1.8
Flavor, color		
Water to make 100% (liquid type)		

Source: Anon. (1966).
[1] Mono-, diglycerides 60%
Sorbitan monostearate 20%   0.3–0.5%
Polysorbate 60 20%
   or
Mono-, diglycerides 75%   0.2–0.4%
Polysorbate 65 25%
[2] Disodium phosphate or sodium citrate may be substituted.

butters, 92° coconut oil, or partially hydrogenated vegetable oils melting below 95°F are recommended for this product.

Ryberg (1968) has given formulations for filled and imitation cream cheese.

Imitation dairy products include coffee whiteners and whipped toppings. Formulations for liquid and spray-dried coffee whiteners are given in Table 29. The various ingredients each have a definite function (Knightly 1969). Miller and Ziemba (1966) recommend that the fat in spray-dried coffee whitener melt at 100°–110°F. Too low a melting point could result in the fat melting during storage with consequent lumping of the powder. Citrate or phosphate salts are added to buffer the protein. This helps prevent feathering when the whitener is added to hot coffee. Coffee contains acid which would coagulate unprotected protein.

TABLE 30

WHIPPED TOPPING FORMULATIONS

| | Weight % | |
Ingredient	Miller and Ziemba (1966)	Thalheimer (1968)
Sugar	7.0	10.0
Corn syrup solids	3.0–5.0	
Sodium caseinate	2.0	2.0
Vegetable gum	0.3–0.5	0.1
Fat	30.0	26.0
Emulsifier	0.35–1.0[1]	1.1[2]
Flavor, color		
Water to make 100%		

[1] Sorbitan monostearate, 60%; polysorbate 60, 40% or glycerol monostearate, 80%; polysorbate 60, 20%.
[2] Mono-, diglycerides, 86%; glyceryl lactopalmitate, 14%.

Two whipped topping formulations are given in Table 30. Both are given in order to illustrate the potential variations in making such product. Thalheimer (1968) points out that if the fat level is less than 30%, more gum is needed to obtain a satisfactory body. He agrees with Miller and Ziemba (1966) that the melting point of the fat should be below body temperature to prevent greasiness in the mouth. Monoglycerides, lactylated monoglycerides, sorbitan esters, and polysorbate esters are the most commonly used emulsifiers. Beck (1967) reports the use of propylene glycol monostearate as an emulsifier for whipped topping. Citrates and phosphates may be used to help stabilize the protein.

The process used for making filled or imitation dairy products is much the same for all types concerned. The milk powder or sodium caseinate is dispersed in cold water in a jacketed, agitated tank along with the sugar and other water soluble solids. The solution is heated to 160°F. Melted fats and emulsifiers are added to the heated aqueous material which is held for 30 min to pasteurize it. The mix is homogenized in a two-stage dairy pressure homogenizer. Pressures used are 1000–1500 psi for the first stage and 500 psi for the second stage. The mix is filled into containers and cooled to 40°F for storage. If the product is to be whipped, it is best to hold the product for 24 hr at 40°F before whipping. This allows the protein to develop optimum whipping properties and whipping stability.

Figure 45 shows a plant layout for preparation of powdered shortening or imitation dairy products. It is essentially the same as one for preparation of any dairy type product except for addition of the spray drying unit and equipment for packaging of dry powder at the end of the process line.

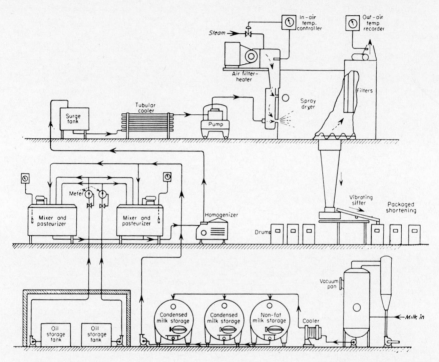

From Alton et al. (1951)

FIG. 45. PLANT LAYOUT FOR MANUFACTURE OF POWDERED SHORTENING

Nonpowdered imitation dairy products are processed similarly but omitting the spray-drying unit.

Reprinted from *Food Engineering*

BIBLIOGRAPHY

ALTON, A. J., NOZNICK, P., and NORTH, G. C. 1951. Piloting a new product? Food Eng. *23*, No. 8, 102–104, 155.

ANON. 1966. Tells how to make coffee whiteners. Food Eng. *38*, No. 2, 135.

BECK, K. M. 1967. Adapt dairy based products to changing consumer patterns. Food Prod. Develop. *1*, No. 2, 34–35.

HETRICK, J. H. 1969. Imitation dairy products, past, present, and future. J. Am. Oil Chemists' Soc. *46*, 58A, 60A, 62A.

KNIGHTLY, W. H.　1969.　The role of ingredients in the formulation of coffee whiteners.　Food Technol. *23*, 171–173, 177, 180, 182.

LOTER, I.　1967.　Process of making chemically acidified sour cream type products.　U.S. Pat. 3,355,298.　Nov. 28.

MILLER, D. E., and ZIEMBA, J. V.　1966.　Build dairy foods to order.　Food Eng. *38*, No. 8, 97–98, 105–106.

RYBERG, J. R.　1968.　Choosing a fat system for filled or imitation dairy cheese.　Food Prod. Develop. *2*, No. 4, 60, 62, 64, 66.

THALHEIMER, W. G.　1968.　Whipped topping, a complex emulsion.　Food Eng. *40*, No. 5, 112–113.

Index

Acetic acid, 44, *See also* Vinegar
  in, mayonnaise, 145
    salad dressing, 172
Acetoglycerides, 44–45
Active oxygen method (AOM), 21, 73–74
Aeration, shortening, 64
Aluminum container, margarine, 139
Aluminum stearate, in pan grease, 105
Annatto, 77, 130
Antifoam agents, *See* Silicones
Antioxidants, 73–74
  in fried foods, 120–121
Antistaling agents, 102

BHA, BHT, 73
  meat fat stabilization, 127
  pigment stabilization, 77
  potato chip stabilization, 120
  safflower oil stabilization, 128
Bixin, *See* Annatto
Bleaching, 51–53
Bread, 102–103, 105
Breading, fried foods, 122
Butter, in chocolate, 195, 197
  in margarine, 131, 134, 140–141
  SFI values, 12
Butyric acid, 78, 140, 198

Cake baking, 68, 70, 92
Cakes, chiffon, 101
  dry mix, 100–101
  standard, 99–100
Carbon dioxide, in mayonnaise and salad
    dressing packaging, 145, 151,
    172
Carotene, heat bleaching, 52
  in palm oil, 37
  pigment, 77
Cellulose, microcrystalline, 174–175
Chlorophyll, 24
  in vegetable oils, 128
Chocolate, 194–207
  coating for frozen dessert, 204
  emulsifiers, 70–71
  enrobing, 202–204
    seed, 202
  evaluation of quality, 204–207
  fat bloom, 196, 198, 204–205
  fineness of grind gage, 206
  flavor and odor, 197
  formulation, 198–199
  gloss, 198, 204, 208
  ingredients, 195–199

liquor, 195–196, 199, 201
manufacture, equipment, 199–201
  process, 201–202
particle size, 206–207
Standards of Identity, 194–195
sugar bloom, 205
viscosity, 198, 209
white, 211
Citric acid, and monoglycerides, 71
  in, margarine, 131
    mayonnaise, 145, 149
    salad dressing, 172
  metal sequestrant, 58, 73–74
Citrus seed oil, 128
Cocoa beans, 195
Cocoa butter, 38–39
  cooling curves, 205–206
  in chocolate, 195–196, 199, 201, 207,
    209
  melting curves, 206–207
  SFI values, 12
Cocoa, Dutch process, 196
  powder, 195–196
    as antioxidant, 196
    in cocoa butter, 38
Coconut oil, 35–36
  in, frying fat, 116, 121
    hard butter, 198, 208–209
  on crackers, 104
  popcorn oil, 124
  SFI values, 12
Coffee whitener, 215
Cold test, 13
  and crystal inhibitors, 76
Colloid mill, 86
Color, 23–24
  artificial, household shortening, 127
    pan and grill shortening, 116
  bleaching, 51–53
  fried food, 113, 122–123
  oil, deterioration, 112, 115
  reversion, 53
Compound (cocoa) coatings, 195, 199,
    207–211
Conche, 199–202
Congeal point, 12
Consistency, shortenings, 18–19
Consistometer bloom, 18–19, 191
Cookie fillers, 106
Cookies, 104–105
Cooling curves, cocoa butter, 205–206
  hard butter, 206
  lard, 16–17, 61